Blue Covenant

THE GLOBAL WATER CRISIS

AND THE COMING BATTLE

FOR THE RIGHT TO WATER

Maude Barlow

THE NEW PRESS

NEW YORK
LONDON

First published in Canada by McClelland & Stewart Ltd., 2007
Published in the United States by The New Press, New York, 2009
Distributed by Perseus Distribution

ISBN 978-1-59558-453-3 (pbk.)

LIBRARY OF CONGRESS CATALOGING-IN-PUBLICATION DATA
  Blue covenant : the global water crisis and the coming battle for the right
to water / Maude Barlow
    p. cm.
  Includes index.
  ISBN 978-1-59558-186-0 (hc.)
    1. Water resources development.   2. Water resources development—
Government policy   3. Water-supply—Political aspects.   4. Water-
supply—Economic aspects.   I. Title
HD1691.B366   2007
333.91'115—dc22   2007037607

The New Press was established in 1990 as a not-for-profit alternative
to large, commercial publishing houses currently dominating the book
publishing industry. The New Press operates in the public interest rather
than for private gain, and is committed to publishing, in innovative ways,
works of educational, cultural, and community value that are often deemed
insufficiently profitable.

www.thenewpress.com

A Caravan book.
For more information, visit www.caravanbooks.org.

Composition by M&S, Toronto
This book was set in Janson

Printed in the United States of America

10   9   8   7   6   5   4   3   2   1

To all the water warriors. You amaze me.

The rule of no realm is mine. . . . But all worthy things that are in peril as the world now stands, those are my care. And for my part, I shall not wholly fail of my task, . . . if anything passes through this night that can still grow fair or bear fruit and flower again in the days to come. For I also am a steward. Did you not know?

– J.R.R. Tolkien

Acknowledgments

Too many people have been incredibly helpful to me in the writing of this book to name them all here. I reference them either in the body of the book or in the Sources and Further Reading section at the end. With all my heart I thank the many activists, academics and friends in the global water justice movement who have advised and counseled me and passed on great research and important stories.

I would especially like to thank all the wonderful people with whom I work at the Council of Canadians and the Blue Planet Project who share my passion for this work on a daily basis. In particular, Anil Naidoo, Susan Howatt, Steven Shrybman, Melanie O'Dell, Meera Karunananthan, Brent Patterson, Brant Thompson, Stuart Trew and Wenonah Hauter were enormously supportive of this project.

As well, I wish to acknowledge and thank both the Canada Council for the Arts for its generous support for this book and the Lannan Foundation, which honored me with the 2005–06 Cultural Freedom Fellowship and a generous grant for my ongoing work.

I have been twice blessed with editors Joel Ariaratnam of The New Press and Susan Renouf of McClelland & Stewart who loved the book from day one and have made it infinitely better than I could have done alone. I am also grateful to Heather Sangster, my copyeditor, and to Elizabeth Kribs at McClelland & Stewart who shepherded this book through a breathtakingly tight schedule.

Finally, I thank my husband, Andrew, who is never-endingly supportive of my work and my passion for the right to water. He and our grandchildren Maddy, Ellie, Angus and Max keep me grounded and give me hope.

Introduction

"Water may also be good for the heart . . ."

– Antoine de Saint-Exupéry, *The Little Prince*

"Suddenly it is so clear: the world is running out of fresh water."

Those were the opening words of my 2002 book, *Blue Gold: The Fight to Stop the Corporate Theft of the World's Water* (co-written with Tony Clarke), which warned that a mighty contest over the world's dwindling freshwater supplies was brewing. Water would become the oil of the twenty-first century, we wrote, and a "water cartel" would develop that would attempt to lay claim to the world's freshwater resources for profit. This would lead to a backlash in communities around the world, we predicted, as well as the growth of a new movement to claim water as part of the global commons.

In the five years since *Blue Gold* was first published, that contest has blown wide open. On one side are powerful private interests, transnational water and food corporations, most First World governments and most of the major international institutions – including the World Bank, the International Monetary Fund, the World Trade Organization, the World Water Council and parts of the United Nations (UN). For these forces, water is a commodity to be sold and traded on the open market. They have established an elaborate infrastructure to promote the private control of water, and they work in close tandem with one another. Their story is told here.

On the other side is a large global water justice movement made up of environmentalists, human rights activists, indigenous and women's groups, small farmers, peasants and thousands of grassroots communities fighting for control of their local water sources. Members of this movement believe that water is the common heritage of all humans and other species, as well as a public trust that must not be appropriated for personal profit or denied to anyone because of inability to pay. Although they lack the financial clout of the water cartel, these groups have found one another through innovative networking and have become a formidable political force on the global scene. Their story is told here as well.

Flush with a number of important recent successes, the global water justice movement has now come together around a common goal: to have water declared, once and for all, a human right, and to have this acknowledgment enshrined at all levels of government, from local ordinances, to nation-state constitutions, to a full UN covenant. The fact that water is not at present a recognized human right has allowed decision-making over water policy to shift from the UN and governments toward global institutions and private interests that favor the big water companies and their commodification of the world's water. This has left untold millions with no legal or moral foundation on which to base their claim that they have the right to enough water for life. Simply put: life requires access to clean water; to deny the right to water is to deny the right to life. The fight for the right to water is an idea whose time has come. It has become the rallying cry of the water justice movement. This story, by no means over, is also told in these pages.

On a personal note, I have been privileged to be an integral part of this incredible struggle, which has taken me to every continent and into remote and often wretchedly poor communities around the globe. It has also taken me deep inside the global institutions and halls of power, where I, along with many others, have confronted the water cartel's determined efforts to control global water policy far into the future.

These journeys have meant so much to me. I offer them to you here in the hope that they will move and inspire you to become a water warrior with us.

Chapter 1

Where Has All the Water Gone?

The Laws of Ecology

*All things are interconnected. Everything goes somewhere.
There is no such thing as a free lunch. Nature bats last.*

– Ernest Callenbach

Three scenarios collude toward disaster.

Scenario one: The world is running out of freshwater. It is not just a question of finding the money to hook up the two billion people living in water-stressed regions of our world. Humanity is polluting, diverting and depleting the Earth's finite water resources at a dangerous and steadily increasing rate. The abuse and displacement of water is the ground-level equivalent of greenhouse gas emissions, and likely as great a cause of climate change.

Scenario two: Every day more and more people are living without access to clean water. As the ecological crisis deepens, so too does the human crisis. More children are killed by dirty water than by war, malaria, HIV/AIDS and traffic accidents combined. The global water crisis has become a most powerful symbol of the growing inequality in our world. While the wealthy enjoy boutique water at any time, millions of poor people have access only to contaminated water from local rivers and wells.

Scenario three: A powerful corporate water cartel has emerged to seize control of every aspect of water for its own profit. Corporations deliver drinking water and take away wastewater; corporations put massive amounts of water in plastic bottles and sell it to us as at exorbitant prices; corporations are building sophisticated new technologies to recycle our dirty water and sell it back to us; corporations extract and move water by huge pipelines from watersheds and aquifers to sell to big cities and industries; corporations buy, store and trade water on the open market, like running shoes. Most importantly, corporations want governments to deregulate the water sector and allow the market to set water policy. Every day, they get closer to that goal. Scenario three deepens the crises now unfolding in scenarios one and two.

Imagine a world in twenty years in which no substantive progress has been made to provide basic water services in the Third World; or to create laws to protect source water and force industry and industrial agriculture to stop polluting water systems; or to curb the mass movement of water by pipeline, tanker and other diversions, which will have created huge new swaths of desert.

Desalination plants will ring the world's oceans, many of them run by nuclear power; corporate-controlled nanotechnology will clean up sewage water and sell it to private utilities, which will in turn sell it back to us at a huge profit; the rich will drink only bottled water found in the few remaining uncontaminated parts of the world, or sucked from the clouds by corporate-controlled machines, while the poor will die in increasing numbers from a lack of water.

This is not science fiction. This is where the world is headed unless we change course – a moral and ecological imperative.

But first we must come to terms with the dimension of the crisis.

We Are Running Out of Freshwater

In the first seven years of the new millennium, more studies, reports and books on the global water crisis have been published than in all of the preceding century. Almost every country has undertaken research to ascertain its water wealth and threats to its aquatic systems. Universities around the world are setting up departments or cross-departmental disciplines to study the effects of water shortages. Dozens of books have been written on all aspects of the crisis. The WorldWatch Institute has declared: "Water scarcity may be the most underappreciated global environmental challenge of our time."

From these substantial and recent undertakings, the verdict is in and irrefutable: the world is facing a water crisis due to pollution, climate change and a surging population growth of such magnitude that close to two billion people now live in water-stressed regions of the planet. Further, unless we change our ways, by the year 2025, two-thirds of the world's population will face water scarcity. The global population tripled in the twentieth century, but water consumption went up sevenfold. By 2050, after we add another three billion to the population, humans will need an 80 percent increase in water supplies just to feed ourselves. No one knows where this water is going to come from.

Scientists call them "hot stains" – the parts of the Earth now running out of potable water. They include Northern China, large areas of Asia and Africa, the Middle East, Australia, the Midwestern United States and sections of South America and Mexico.

The worst examples in terms of the effect on people are, of course, those areas of the world with large populations and insufficient resources to provide sanitation. Two-fifths of the world's people lack access to proper sanitation, which has led to massive outbreaks of waterborne diseases. Half of the world's hospital beds are occupied by people with an easily preventable waterborne disease, and the World Health Organization reports that contaminated water is implicated in 80 percent of all sickness and disease worldwide. In the last decade, the number of children killed by diarrhea exceeded the number of people killed in all armed conflicts since the Second World War. Every eight seconds, a child dies from drinking dirty water.

Some wealthier countries are just beginning to understand the depth of their own crisis, having adopted a model of unlimited consumer growth based on industrial, trade and farming practices that are wasting precious and irreplaceable water resources. Australia, the driest continent on Earth, is facing a severe shortage of water in all of its major cities, as well as widespread drought in its rural countryside. Annual rainfall is declining; salinity and desertification are spreading rapidly; rivers are being drained at an unsustainable rate; and more than one-quarter of all surface water management areas now exceed sustainable limits. Climate change is accelerating drought and causing freak storms and weather patterns just as the population is set to expand dramatically in the next twenty years. (Ironically, this is, in part, to take in the climate-change refugees such as the inhabitants of the Solomon Islands, who will lose their lands to the rising seas.)

Many parts of the United States are also experiencing severe water shortages. Pressure is mounting on the Great Lakes governors to open up access to the lakes to the burgeoning megacities around the basin. In 2007, Lake Superior, the world's largest freshwater lake, dropped to its lowest level in eighty years and the water has receded more than fifteen meters from the shoreline. Florida is in trouble. The state's burgeoning population, with a net influx of 1,060 people every day, relies almost entirely upon its dwindling groundwater sources for its water supplies. To keep its fast-spreading lawns and golf courses green, the Sunshine State is sucking up groundwater at such a rate that it has created thousands of sinkholes that devour anything – houses, cars and shopping malls – unfortunate enough to be built on them. California has a twenty-year supply of freshwater left. New Mexico has only a ten-year supply. Arizona is out: it now imports all of its drinking water. Lake Powell, the man-made backup for the western water supply, has lost 60 percent of its water. A major June 2004 study by the National Academy of Sciences and the U.S. Geological Survey found that the parched Interior West is probably the driest it has been in five hundred years. As in Australia, anxious American politicians talk about "drought" as if this is a cyclical situation that will right itself. But scientists and water managers throughout the American Midwest and Southwest are saying that it is more than a drought: major parts of the United States are running out of water. In fact, the Environmental Protection Agency warns that if current water use continues unchecked, thirty-six states will suffer water shortages within the next five years.

Because of the wealth of these countries, most of their populations are still not suffering from water shortages. That is not so for those in the global South – hence the term *water apartheid*. The world's poor who are living without water are either in areas that do not have enough water to begin with (Africa), where surface water has become severely polluted (South America, India) or both (Northern China). Most of the world's megacities – those with ten million or more inhabitants – lie within regions experiencing water stress. These include Mexico City, Calcutta, Cairo, Jakarta, Karachi, Beijing, Lagos and Manila.

In 2006, the number of city dwellers surpassed the number of rural dwellers for the first time in history. The urban populations of the Third World are growing exponentially, creating enormous slums without water services. In the last decade, the number of city dwellers without reliable access to clean water increased by more than sixty million. By 2030, says the UN, more than half the population of these huge urban centers will be slum dwellers with no access to water or sanitation services whatsoever. One report cited a current example of an area in Mumbai, where one toilet serves 5,440 people.

Not surprisingly, there is a huge gulf between the First World and the Third World in water use. The average human needs fifty liters of water per day for drinking, cooking and sanitation. The average North American uses almost six hundred liters a day. The average inhabitant of Africa uses six liters per day. A newborn baby in the global North consumes between forty and seventy times more water than a baby in the global South.

These appalling disparities have rightly created a demand for more water equity and a commitment to providing water for the 1.4 billion people currently living without it. The UN Millennium Development Goals include reducing by half the proportion of people living without safe drinking water by the year 2015. While laudable, this initiative is failing not only because the UN has worked with the World Bank to promote a flawed model for water development (see Chapter 2), but also because it assumes that there is enough water for everyone without seriously addressing the massive pollution of surface waters and the consequent massive overmining of groundwater supplies.

Our Surface Waters Are Polluted

We were all taught certain fundamentals about the Earth's hydrologic cycle in grade school. There is a finite amount of available freshwater on the planet, we learned, and it makes its way through a cycle that ensures its safe return to us for our perpetual use. In the hydrologic cycle, water vapor condenses to form clouds. Winds move the clouds across the globe, spreading the water vapor. When the clouds cannot hold the moisture, they release it in the form of rain or snow, which either seeps into the ground to replenish groundwater or runs off into lakes, streams and rivers. (This is the water – less than one-half of 1 percent of all the water on Earth – available for human use that does not deplete the stock.) As these processes are happening, the power of the sun is causing evaporation, changing liquid water into vapor to renew the cycle. About four hundred billion liters of water are cycled through this process every year. In this scenario, the planet could never "run out" of water.

But this cycle, true for so many millennia, did not take into account modern humans' collective capacity for destruction. In the last half-century, the human species has polluted surface waters at an alarming and accelerating rate. The world may not exactly be running out of water, but it is running out of clean water. Ninety percent of wastewater produced in the Third World is discharged, untreated, into local rivers, streams and coastal waters. As well, humans are now using more than half of accessible runoff water, leaving little for the ecosystem or other species.

In China, 80 percent of the major rivers are so degraded they no longer support aquatic life, and an astonishing 90 percent of all groundwater systems under the major cities are contaminated.

China is now home to seven of the ten most polluted cities in the world. The World Health Organization reports that 700 million of the 1.3 billion people of China drink water that doesn't even meet the most basic minimum safety standards set by that world body. In late 2006, the Chinese government reported in a rare admission of failure that, as a result of massive pollution, more than two-thirds of Chinese cities face water shortages, with at least one hundred of them facing immediate depletion. Forty-five billion tons (about forty-one trillion kilograms) of untreated wastewater are pumped directly into lakes and rivers every year, according to a recent article in the *China Daily*.

This scenario is repeated in many parts of Asia. A 2005 nationwide survey in Pakistan revealed that less than 25 percent of the population has access to clean drinking water due to massive pollution of the country's surface waters. The Indonesian Environment Monitor reports that Indonesia has one of the lowest sanitation rates in the world. Less than 3 percent of Jakarta's residents are connected to a sewer, leading to severe pollution of nearby rivers and lakes and the contamination of 90 percent of the city's shallow wells. Almost 65 percent of Bangladesh's groundwater is contaminated, with at least 1.2 million Bangladeshis exposed to arsenic poisoning.

Seventy-five percent of India's rivers and lakes are so polluted, they should not be used for drinking or bathing. More than 700 million Indians – two-thirds of the population – do not have adequate sanitation, and 2.1 million Indian children under the age of five die every year from dirty water. The fabled Yamuna River is clinically dead, killed as it makes its way through New Delhi's teaming slums. The coasts of Mumbai, Madras and Calcutta are putrid. The sacred Ganges, where millions come to worship, is an open sewer. Thousands of Hindu worshippers boycotted the 2007 religious festivals in which millions plunged into the Ganges to wash away their sins. One Indian government study called the situation in India "an unparalleled water crisis." Against this backdrop of pollution and scarcity, India's urban water demand is expected to double by 2025, and industrial water demand will triple.

The statistics for Russia are hauntingly similar. The U.S. Library of Congress reports on water pollution in Russia – a phenomenon little reported in Russia itself. Seventy-five percent of Russia's inland surface water is polluted and approximately 30 percent of the groundwater available for use is highly polluted. Many rivers are carriers of waterborne killers, and 60 percent of rural residents are drinking water from contaminated wells.

The underground reservoir of the Mountain Aquifer is the most important source of water for Israelis and Palestinians, supplying more good-quality water per year than any source between the Mediterranean Sea and the Jordan River. But, reports Friends of the Earth Middle East, the sewage of more than two million people who live above the aquifer is discharged untreated into streams and other natural water sources percolating into the groundwater. This amounts to almost sixty-one million cubic meters a year.

According to the European Commission, 20 percent of all surface water in Europe is "seriously threatened," and the UN adds that only five of the fifty-five major rivers in Europe can be considered "pristine" anymore. Belgium's water is singled out as particularly bad, due to heavy pollution by industry. The Rhine, the Sarno and the Danube rivers are all in peril. Recent and regular droughts have European leaders very worried about water availability. Southern Spain, southeastern England and western and southern France are all viewed as chronically vulnerable, while fears are growing in Portugal, Italy and Greece. In May 2007, a state of emergency was declared in the northern and central regions of Italy as the country's largest river, the Po, dried up, devastating the Po Valley, which grows a third of the country's food. In several of these countries, reservoirs are at their lowest levels in recorded history.

Forty percent of U.S. rivers and streams are too dangerous for fishing, swimming or drinking, as are 46 percent of lakes due to massive toxic runoff from industrial farms, intensive livestock operations and the more than one billion pounds of industrial weed killer used throughout the country every year. Two-thirds of U.S. estuaries and bays are moderately or severely degraded. The Mississippi River carries an estimated 1.5 million metric tons of nitrogen pollution into the Gulf of Mexico every year. Every year, one-quarter of the U.S. beaches are under advisories or closed due to water pollution. The U.S. government refuses to ban the herbicide Atrazine, an endocrine disrupter banned in many countries around the world and widely linked to cancer. In Canada, more than one trillion liters of untreated sewage is dumped into waterways each year, a volume that would cover the entire 7,800-kilometer length of the Trans-Canada Highway, six stories high.

In Latin America and the Caribbean, more than 130 million people do not have safe drinking water, and only 86 million (of 550 million) are connected to adequate sanitation systems. Seventy-five percent of the population suffers from chronic dehydration because of poor water quality. Basic drinking water and sanitation are out of the reach of a third of Peru's urban dwellers and two-thirds of its rural populations. Major cities such as Mexico City and São Paulo are facing the twin threats of massive overconsumption of water and mass contamination. Less than 10 percent of Mexico City's waste is recycled in a city of more than twenty million people. And that is higher than the average: only about 2 percent of Latin America's wastewater receives any treatment at all.

More than one-third of Africa's population currently lacks access to safe drinking water, and within fifteen years, one in two Africans will be living in countries that are confronted with serious water stress. Of the twenty-five countries in the world whose people have the least access to safe, clean water, nineteen are in Africa. Lake Victoria, the source of the Nile, is being used as an open sewer. It and dozens of other African lakes and rivers are imperiled, according to the UN Environmental Program, whose October 2005 report, *The Atlas of African Lakes*, used satellite images to reveal the unprecedented deterioration of all of Africa's 677 major lakes. As well, the report reveals alarming drops in water levels in most of Africa's lakes. Lake Chad has shrunk by almost 90 percent.

Thousands of Angolans died in a 2006 cholera outbreak caused by filthy water. Only one in six Luandan households have basic sanitation services, and the city's 4.5 million residents live in the midst of mountains of garbage and open sewers in the streets. Eighty percent of South Africa's rivers are imperiled by pollution, and every year, residents (usually women) have to walk farther and farther to find clean water. The women of South Africa now collectively walk the equivalent of the distance to the moon and back sixteen times a day for water.

One inevitable result of the massive pollution of surface waters in poor countries is that sewage water is increasingly being used to fertilize crops. In 2004, the Sri Lankan bureau of the International Water Management Institute undertook the first global survey on the hidden practice of wastewater irrigation. It found that fully one-tenth of the world's irrigated crops – from lettuce and tomatoes to mangoes and coconuts – is watered by sewage, most of it completely untreated, "gushing direct from sewer pipes into fields at the fringes of the developing world's great megacities." The sewage is added to fields complete with disease-causing pathogens and toxic waste from industry. In some Third World metropolises, all food sold is grown in sewage.

Our Groundwater Sources Are Depleting

To deal with this vast pollution and the resulting effect of reduced clean water supplies, farms, cities and industries all over the world are turning to groundwater sources, using sophisticated technology to drill deep into the Earth and pull up ancient aquifer water for daily use. This is a second piece of the "running out" puzzle. We are taking water from where it is accessible – in aquifers and other groundwater sources – and putting it where it gets used and lost, such as in mass irrigation of deserts, to make cars and computers, or to produce oil from tar sands and coal methane beds where it becomes polluted or actually lost to the hydrologic cycle.

The current practice of "water mining" is different from the sustainable use of well water that has served farmers for generations. Today, groundwater is seen as a finite resource such as a mineral – a deposit to be mined until it is gone, allowing the searcher to move on to new sites – rather than a renewable resource that must be managed and replenished. The exponential mining of groundwater is largely unregulated, and no one knows when the limit will be hit and the supply depleted within a certain community or region.

We do know that the use of groundwater for daily living is growing very fast. About two billion people – one-third of the world's population – depend on groundwater supplies, withdrawing approximately 20 percent of global water annually. Groundwater aquifers are being overpumped almost everywhere in the world and are also being polluted with chemical runoff from industrial farming and mine tailings, as well as being invaded by saltwater from careless drilling practices. (In some cases, overextraction of a river exposes an aquifer to danger. The Dead Sea is disappearing, victim to widespread abuse of the waters of the Jordan River for irrigation. As the Dead Sea withdraws, aquifers surrounding it are left at a higher level than its surface. Underground water flows into the sea, drying out aquifers that have been untouched for millions of years.)

In the First World, much of the groundwater extraction is due to big industrial agribusiness taking massive amounts of water using huge industrial bores. In the Third World, the problem is caused by millions and millions of small farmers using personal pumps.

Groundwater mining can be traced in great part to the famous Green Revolution and the use of flood irrigation to mass-produce food. Since 1950, the global acreage of land under irrigation – the driving factor behind the Green Revolution – has tripled. Using vast amounts of water, scientists developed high-yield crop varieties to meet the needs of developing nations. While the "revolution" produced more food, it used way too much water and also depended on copious amounts of dangerous pesticides and fertilizers. Some countries abandoned past sustainable farming practices and started "double cropping," whereby crops are grown during the dry season *and* the wet season, adding to the demands on water.

As British environmentalist Fred Pearce points out, irrigated farming gave us twice as much food but used three times as much water, and did more harm than good. He lists the world's major rivers that no longer reach the sea: the Colorado and the Rio Grande in the United States; the Nile in Egypt; the Yellow River in China; the Indus in Pakistan; the Murray in Australia; the Jordan in the Middle East; and the Oxus in Central Asia. They have been depleted from damming, overuse and the mining of the groundwaters that feed them.

In *Pillar of Sand*, about the growing desertification of the planet, Sandra Postel argues that the changes in food production over the last fifty years have put a profound strain on the world's groundwater supplies. The agricultural practices of many countries are sustained by the hydrological equivalent of deficit financing. At least 10 percent of the global grain harvest is grown with groundwater supplies that are not being replenished, an amount equal to the total flow of two Nile Rivers every year.

Groundwater extraction often turns oases into deserts, but it can also literally turn a desert into an oasis. The Ogallala Aquifer is a vast geologic formation that sprawls underneath eight states from South Dakota to Texas. Early settlers in the semi-arid High Plains were plagued by crop failure due to cycles of drought, culminating in the dust bowl of the 1930s. After the Second World War, the technology was developed to mine the Ogallala, and the High Plains was turned into one of the most agriculturally productive regions in the world. Its massive water reserves – larger than Lake Huron – are now used to grow water-intensive crops such as cotton and alfalfa in the desert. But the miracle will not last. Because it is so deep, the Ogallala gets very little replenishment from nature to offset the two hundred thousand borewells working 24-7 to remove its ancient treasure. In several short decades, it has lost – forever – a volume of water equivalent to the annual flow of eighteen Colorado rivers. It is now producing half as many crops as it did in the 1970s, but demand continues to grow.

This story is repeated all over the United States, which is now dependent on nonrenewable groundwater for an astonishing 50 percent of its daily water. Groundwater supplies Europe with 65 percent of its drinking water, and the European Commission warns that 60 percent of European cities exploit their groundwater resources. Half of Europe's wetlands are endangered from groundwater mining, and the groundwater itself is becoming seriously polluted. Aquifers are way overpumped in Australia as well – groundwater extraction skyrocketed a whopping 90 percent in the 1990s – and is contaminated from the eighty thousand toxic dumpsites under Australia's major cities.

However, it is in Asia that the coming crisis can be seen most clearly. The London-based *New Scientist* reported scientists' findings on what it called a "little-heralded crisis" all over Asia with the unregulated and exponential drilling of groundwater. Farmers are drilling millions of pump-operated wells in an ever-deeper search for water and are threatening to suck the continent's underground reserves dry, setting the stage for "untold anarchy." Vietnam has quadrupled its number of tube wells in the past decade to one million, and water tables are plunging in the Pakistani state of Punjab, which produces 90 percent of the country's food.

In India, twenty-three million tube wells operate around the clock using technology borrowed from the oil industry, going so deep they are taking up water formed at the time of the dinosaurs. Every year another million wells are added. The pumps are taking two hundred cubic kilometers of water out of the Earth every year, with only a fraction of that replaced by monsoon rains. (A cubic kilometer is the volume of water equal to that of a cube of one kilometer each side.) The farmers are forced to drill deeper to keep pace with falling water tables, and thousands of farmers have committed suicide in the last decade when their farms ran out of water completely. Plunging water tables in Tamil Nadu and northern Gujarat have cut the land available for farming in half. This situation will be repeated all over India, say water experts there.

China has less water than Canada and forty times more people. In Northern China, groundwater depletion has reached catastrophic levels. Across the northern half of the country – China's breadbasket – groundwater pumping amounts to some thirty billion cubic meters a year. This is due to massive over-pumping for agriculture but also because government planners divert large amounts of water from farming to industry every year, to fuel China's economic "miracle." The water table beneath Beijing has fallen nearly two hundred feet in the past twenty years, which has led some planners to warn that China may have to choose another city for its capital.

Drought-related sandstorms are already plaguing China. In the first half of 2006, thirteen major sandstorms had hit Northern China. In April 2006, one storm swept across an eighth of the country and even reached Korea and Japan. On its way, it dumped a mind-boggling 336,000 tons (about 305 million kilograms) of dust on Beijing, forcing people to walk around with facemasks for protection. Every year, a new desert the size of Rhode Island is created in China.

The Planet Is Drying Up

Melting Glaciers

China's crisis is exacerbated by the rapid melting of the Tibetan glaciers, which are vanishing so fast (due to climate change) they will be reduced by 50 percent every decade, according to the Chinese Academy of Sciences. Each year, enough water melts from the 46,298 glaciers of the plateau to fill the entire Yellow River. But rather than adding freshwater resources to a thirsty country, the furious pace of this melting is actually creating deser-tification. Instead of steadily feeding the great rivers of Asia – the Yangtze, the Indus, the Ganges, the Brahmaputra, the Mekong and the Yellow – as the Himalayan glaciers have done for millen-nia, the fast-melting water running off the plateau increases soil erosion, allowing the deserts to spread, and then evaporates before it reaches the thirsty rivers.

Says the Academy's Yao Tandong, "The full-scale glacier shrinkage in the plateau regions will eventually lead to an ecological catastrophe." The Indus, for example, provides water for 90 percent of Pakistan's crops.

The World Wildlife Fund (WWF) echoes the concern, reporting that billions of people worldwide face severe water shortage as the world's glaciers experience meltdown. The WWF singles out Ecuador, Peru and Bolivia for concern, as all are dependent on the glacier melt from the Andes for water supplies. In 1980, 75 percent of the European alpine glaciers were advancing. Today, 90 percent are in retreat. The Swiss Alps – the major source of water for the Rhine, the Rhone and the Po rivers – are melting twice as fast as any other in the world. In Canada, the glacier that feeds Alberta's Bow River is melting so quickly that in fifty years, there will not likely be any water left in the river except for the occasional flash flood.

The state of the world's mountains – the source of half of humanity's drinking water, now referred to by scientists and environmentalists as "water towers" – should be a major concern to us all as global warming strips away their ancient glaciers. Receding glaciers at sea are another loss of freshwater, as they melt into saltwater and add to the rising of the oceans. Glacier melt is yet another piece of the "running out" puzzle – another example, such as groundwater mining, where water is removed from where it has been stored for millennia to provide life for humans and nature, and ends up lost to both.

Virtual Water Trade

Water is also massively displaced through the trade in what is called *virtual water*, a term that describes the water used in the production of crops or manufactured goods that are then exported. Israeli economists first used the term *virtual* or *embedded* water in the early 1990s when they realized that it didn't make sense from an economic point of view to export scarce Israeli water. This is what was happening, they said, every time water-intensive oranges or avocados were exported from their semi-arid country. Because of poor water management systems used around the world (more than half of all water used in flood irrigation is lost to seepage or evaporation), even a small bag of salad takes three hundred liters of water to produce. It takes about a thousand liters to produce a kilogram of wheat and five to ten times as much to produce a kilogram of meat. Up to thirty thousand liters of water are used to produce one kilogram of cotton.

Water that is used in the production of food is "virtual" because it is not contained anymore in the product, even though a great deal of it was used in the production process. If a country exports a water-intensive product to another country, it amounts to exporting water in a virtual form, although no water is technically being traded or sold. This diminishes the amount of water consumed in the importing country. Wealthy countries with low water supplies, such as Saudi Arabia and Netherlands, import much of their water through food imports from countries that either have lots of water or who are too poor to have a choice but to exploit what is left of their water. Japan, for instance, imports about 65 percent of the total volume of water that it uses to produce the goods and services consumed by its citizens (this is called the water footprint of a country) through the import of crops and goods that use other countries' water for production. For some water-rich countries such as Canada, this practice may appear to be benign. But many poor countries are exporting huge amounts of water through virtual water trade because of a desperate need for income and because they have been strongly pushed by the World Bank and the International Monetary Fund to pay off their debts through monoculture crop exports, even if it means using their best, most arable land and their remaining water supplies to do so.

India, with its water crisis on the doorstep, is a major virtual water exporter, as is Thailand. Vietnam is destroying its water table to grow coffee for export. Africa supplies much of Europe with out-of-season fruits and vegetables, just as Latin America provides for North America. Kenya is destroying the waters of Lake Naivasha to grow roses for export to Europe. Scientists predict the lake, the source of water for Africa's largest population of hippopotamuses, will be a "putrid muddy puddle" within five to ten years if its draining for flower irrigation is not halted. (Knowing this, the big European flower companies are already planning to relocate to Ethiopia and Uganda.)

As well, many developing countries are growing "biofuels" – energy replacements derived from sugarcane, corn, palm oil and soy – to meet demands in global Northern countries for alternatives to oil and gas. Biofuels – food to feed cars – have come under intense criticism not only because they take up vast areas of agricultural land to cultivate and are energy-intensive crops in themselves – thus exporting the energy costs of the North to the South – but because they use massive amounts of water. As agricultural sciences professor David Pimentel of Cornell University reports, it takes seventeen hundred liters of water to produce one liter of ethanol, when the water used to process the corn to biofuels is added to the water used to grow the corn, usually using wasteful flood irrigation practices. China imports about twenty million tons (about eighteen billion kilograms) of soy biofuel every year, mostly from Brazil (a factor perhaps in the biofuel pact that was signed by Presidents Bush and Lula in the American President's March 2007 visit to Brazil). For this, the producing countries use forty-five cubic kilometers of water – about half the annual domestic consumption of water for the whole world. In northern Brazil, where the big biofuel plantations are numerous, whole rivers are drying up. (Not all biofuels are for export. The Canadian and U.S. governments are also promoting the growth of biofuels in their agriculture sectors with large subsidies. *The Sacramento Bee* estimates that, to meet California's stated goal of ethanol production, the state will have to find ten trillion liters of extra water a year.)

Many poor countries are exporting their way to drought. Between 15 and 20 percent of the water used in the world for human purposes is not for domestic consumption but for export, according to the UN, in what is considered by many to be a conservative estimate. But with the continued emphasis of the World Bank and other global financial institutions on export growth, this practice is bound to increase and, with it, the transfer of water from the poor to the rich. Oddly, two wealthy but water-stressed countries are also major virtual water exporters: the United States and Australia. Net exports of water from the United States amount to one-third of the total water withdrawal in that country and are a major factor in the drying out of the American Midwest and Southwest. Not coincidentally, both Australia and the United States have governments in denial about their water crises and completely wedded to economic globalization and its false promise of unlimited growth.

Urbanization and Deforestation

Yet another answer to the question of where the world's water has gone is that it is displaced from the hydrologic cycle by massive urbanization and paving over of natural environments. In a groundbreaking study, Slovakian scientist and Goldman Prize–winner Michal Kravçik showed that when water cannot return to fields, meadows, wetlands and streams because of urban sprawl and the removal of green spaces, there is less water in the soil and local water systems and, therefore, less water to evaporate from land. It is as if the rain is falling on a large cement umbrella, which carries it out to sea. The destruction of water-retentive landscapes means that less precipitation remains in river basins and continental watersheds; this in turn equates to less water in the hydrologic cycle.

Kravçik chronicles the death of many other societies in the past from the very water-destructive practices we use widely today. He explains that water is a thermal regulator that moderates weather extremes. The more water the atmosphere has, the stronger will be the moderating effects on temperature and weather. Most evaporating water in the closed hydrologic cycle condenses again in the local watershed. There, it needs plentiful vegetation for the process of "transpiration" – the process whereby plants and trees "sweat" water, cooling the air in the process. If the water cycle is disrupted because the vegetation has been removed, water vapor is lost to the local watershed. The elimination of vegetation from the soil by removing forests, over-grazing land or using poor farming methods was a major cause of the downfall of past civilizations. Modern humans have added urbanization and the practice of removing huge amounts of fresh-water through sewage systems, many of which dump freshwater directly into the oceans. Kravçik says the destruction of vegeta-tion, combined with the directing of rainwater from continents into the oceans, is as great a cause of global warming and rising seas as greenhouse gas emissions.

An added problem is the creation of urban heat islands that are warmer than the surrounding rural areas. As *Science News* reported, "impervious surfaces" the size of Ohio now cover the United States and actually affect the local climate. When precip-itation doesn't soak into urban landscapes, it isn't available to absorb heat, evaporate and thus cool the environment. Cities lose their capacity to "sweat."

The problem is exacerbated by deforestation. In a March 2005 study by the Australian Nuclear Science and Technology Organization, scientists analyzed variations in the molecular structure of rain along the Amazon River. This allowed them to "tag" the water as it flowed into the Atlantic, evaporated, blew back inland to fall again as rain and finally returned to the river. The study showed that since the 1970s, when intensive deforest-ation began, the ratio of the heavy molecules found in the rain over the Amazon had declined significantly. The only possible explanation was that the molecules were no longer being returned to the atmosphere to fall again as rain because the veg-etation was disappearing. The team found a clear connection between the degraded forest and reduced rainfall – an association with a long anecdotal history but lacking in scientific proof until the Australian study.

Desertification and Climate Change

This drying trend has recently been verified by a number of important sources. The U.S. National Center for Atmospheric Research (NCAR) reports that the percentage of the Earth's land area stricken by serious drought more than doubled between the 1970s and 2005. Widespread drying occurred over much of Europe, Asia, Canada, western and southern Africa and eastern Australia. In Nigeria, two thousand square kilometers is becoming desert every year.

As well, researchers from the NASA-sponsored Gravity Recovery and Climate Experiment (GRACE) are using a pair of roving satellites to measure changes in the water supply around the world. The two satellites measure the gravitational field of the Earth; minute changes in the data can then be extrapolated to show where water is "displaced," even if it is captured in snow, rivers or aquifers. Although the project is relatively new (launched in 2003), it has already identified the central valley of California, parts of India and large areas of Africa for special concern. The annual 21.6-millimeter shrinkage in the depth of the Congo translates to 260 cubic kilometers, or roughly the annual flow of fourteen Colorado rivers. *Every year.*

A major October 2006 report from the U.K. Meteorological Office reproduced total global water trends over the last fifty years and then applied the model to forecast the future. The study clearly showed that the current extent of drought could double by the end of the twenty-first century, threatening the survival of millions of people around the world. In contrast, in the second half of the last century, just 1 percent of the world was affected by extreme drought.

There are several ways in which climate change affects freshwater sources. As seas rise, they will take out more wetlands, which are already under siege. Wetlands have been called the kidneys of freshwater systems, as they filter and purify dirt and toxins before they reach rivers, lakes and aquifers. (Forests are the lungs of the water system, absorbing pollution and preventing flooding.) Further, as global warming raises the temperature of the Earth, the soil water needed to sustain the freshwater cycle will evaporate more readily. Water in lakes and rivers will evaporate more quickly as well, and the snow packs and ice cover that replenish these systems will become more rare.

Climate change and reduced water in the hydrologic cycle will create as many as one billion climate-change refugees, many of them from water loss, warned the development agency Christian Aid in a May 2007 report titled *Human Tide: The Real Migration Crisis*. Citing research by Oxford academic Norman Myers, which concluded that by 2050, five times as much land is likely to be under "extreme" drought as today, another Christian aid group, Tearfund, has called on world leaders to get beyond rhetoric. In its report *Feeling the Heat*, Sir John Houghton, one of Britain's leading climate scientists, warned that water shortage would be the most visible and terrible climate threat in developing countries.

High Technology Solutions Are Part of the Problem

Where they are taking proactive steps to alleviate the water crisis, many nations and the international financial institutions are promoting the high-technology solutions of dams, diversions and desalination. While it is difficult to imagine a world without these fixtures, in the long run, all are part of the problem and they will not provide the answers we need. On the contrary, these expensive technologies all have the potential to do great harm to the ecosystems in which they are placed, further exacerbating the global water crisis.

Dams

More than forty-five thousand large dams (higher than fifteen meters) have been built around the world at a cost of around US$2 trillion. While dams can provide some benefits, such as generating electricity, supplying water, controlling floods and facilitating navigation, much evidence suggests that these are benefits of smaller dams. Large dams trap organic materials and rotting vegetation from submerged lands, which in turn creates methane gas, a major source of greenhouse gas emissions. They also displace huge numbers of people due to their size. Close to eighty million people have been forced from their lands to make way for dams and few have been compensated. Sixty percent of the world's major rivers have been fragmented by dams and diversions, and more than a million square kilometers – 1 percent of the world's land surface – have been inundated by reservoirs worldwide.

Big dams disrupt river flow patterns and aquatic habitat, reducing biodiversity. Big dams and diversions are the main reason why one-third of the world's freshwater fish species are extinct or endangered, reports the International Rivers Network (IRN). Large dams are also the reason why so many of the world's great rivers no longer reach the ocean, and why rich delta areas, where freshwater meets seawater – the home to so many species – have been destroyed. The World Wildlife Fund reports that only 21 of the world's 177 longest rivers run unhindered to the sea.

Perhaps most importantly, big dams significantly contribute to the emission of greenhouse gases, and therefore to global warming, one of the greatest threats to freshwater resources. Brazilian climate change expert Philip Fearnside estimates that hydropower dams in the Amazon cause far more global warming than modern natural gas plants generating the same amount of energy. "It may seem counterintuitive," says IRN's Patrick McCully, "but tropical hydropower reservoirs can have a far greater impact on global warming than even their dirtiest fossil fuel plant rivals."

Although the bloom is off the rose for big dams in the global North, the World Bank and other regional development banks are planning many new ones in India, China, Brazil, Turkey, Iran, Laos, Vietnam, Mexico and Ethiopia – all areas with severe water shortages and/or pollution.

Diversions

Another high-tech answer to the water crisis is to take water from where it is exists in nature and move it to big cities or industries far away. In the past, water was diverted through canals. Now, however, water is often carried by giant pipes than can take it very far from its source. Increasingly, around the world, a massive network of pipelines is being constructed to move water from place to place, just like the pipeline network that now moves vast quantities of oil and gas. This is being done with no coordinated planning or any understanding of what this might mean ecologically. These pipelines are very expensive and, like energy pipelines, damaging to the environment. They disrupt wildlife and ecosystems and, in colder regions, must be constructed in permafrost.

When water is taken from a watershed, where it is needed as the lifeblood of the ecosystem, the result causes water-level drawdowns in the short-term and can result in full water depletion in the long-term. This scenario is already setting the needs of rural, indigenous and farm communities against large urban centers. It is also causing tensions among nations when one country claims groundwater that is claimed by other countries as well. And it is a leading cause of the desertification of rural areas, whose water systems are sold, expropriated or just plain stolen.

Mexico City is facing a chronic and severe water crisis as its population continues to expand and its water supplies disappear. The state now pipes freshwater – sixteen thousand liters a second – into Mexico City from a reservoir in the Mazahuas indigenous community one hundred kilometers away. The Mazahuas have been waiting for three decades for access to this water, which was confiscated by the state in 1980 (and before then was their exclusive resource), and have promised to take to arms if this injustice is not rectified. Mexican state authorities are going into many other communities surrounding the city, some quite far away, searching for new water supplies for expropriation.

Libya consists mostly of desert terrain, and rapid development of its small coastal areas has put a severe strain on its limited water supply. So in 1980, Colonel Mu'ammar Gadhafi initiated the Great Man-Made River Project to extract water from the aquifers under the Sahara Desert. He built a gigantic five-thousand-kilometer, us$35 billion underground pipeline – the biggest in the world to date – that has made the coastal desert bloom. As well, more than thirteen hundred wells have been drilled into the aquifer, some as deep as five hundred meters. At present, an astonishing 6.5 million cubic meters of water are being diverted every day from the aquifer. This "Eighth Wonder of the World" has two problems, though. First, the waters of this aquifer lie under several other countries, namely Chad, Egypt and Sudan, who also have claims to it. Second, and more importantly, the aquifer has no recharge source and will eventually be completely exhausted.

Israel is considering building a World Bank–funded, two-hundred-kilometer pipeline to carry water from the Red Sea to "replenish" the Dead Sea, which has dramatically shrunk in size in recent years. Environmentalists are warning that, instead of saving the Dead Sea, the project could further damage it by encouraging algae growth.

India is planning to build a massive pipeline from the Tehri Dam in the outer Himalayas (when it is finished, the Tehri will be the fifth largest dam in the world and will submerge forty-two-hundred hectares of fertile farmland) to divert water from the Upper Ganga Canal – the main source for the sacred Ganges – to supply Delhi with drinking water. This is one step in a proposal eventually to link all of India's rivers through the creation of super-dams and massive water diversion through canals and pipelines. The proposed cost of this unprecedented diversion would be two hundred times what India spends on education, and three times what it collects in taxes.

China is setting the stage to rewrite its future through vast engineering feats that involve diverting water cascading from the Tibetan Highlands to the parched Yellow River in the country's west. The Tibetan Western Route of the South-North Water Transfer Project is slated to begin construction as early as 2010. It will join the Central and Eastern Routes, already under construction to draw water from the Yangtze River for Beijing, and will consist of three 1,100-kilometer channels and pipelines, projected to cost US$300 billion. In its first phase, the scheme will transfer about four billion cubic meters of water annually – the size of California's main water transfer scheme – and will eventually deliver 46 trillion liters of water a year.

Russian authorities are up in arms over China's plan to build a 300-kilometer irrigation canal and siphon off 450 million cubic meters of water per year from Siberia's Irtysh River, now shared by the two countries. As many as two million Russians could have their water supply cut off if this project is not halted. As well, rumors continue to persist that a massive pipeline is also in the planning stage to pipe the water of Russia's famed Lake Baikal to China and even, eventually, to the Middle East and the United States. Lake Baikal is the world's largest source of freshwater, larger than all the Great Lakes of North America put together. In August 2005, Russian and Chinese scientists carried out the first joint-research mission to explore the environment around the lake and the quality of the water.

A number of pipelines are being proposed in the developed world as well. The European Commission is backing the establishment of a European Water Network that would divert water from the Austrian Alps by pipeline to thirsty areas of the south of Europe. A plan is underway to build a 650-kilometer pipeline to divert water from the Missouri River to the populated areas of South Dakota, southwestern Minnesota and northwestern Iowa. The Southern Nevada Water Authority is proposing to divert water from southern Nevada to Las Vegas through a five-hundred-kilometer pipeline. Utah has proposed a US$500 million, two-hundred-kilometer pipeline from Lake Powell to serve St. George and Washington County. Many projects have been proposed (and shelved due to stiff opposition) to pipe water from Canada's north to the American Midwest. As water becomes more precious, many are looking once again at these schemes. Former Australian Minister for the Environment and Water Resources Malcolm Turnbull favored a pipeline to carry water from the rivers of New South Wales north to the thirsty cities of Queensland. In an April 2007 report, Turnbull claimed that even with the high cost of pipeline construction, this would provide more water at less cost than desalination plants.

As far back as sixty years ago, water was massively diverted from the Aral Sea through a dredged canal and sent to the desert to grow cotton for export. At the time, the Aral was the world's fourth largest lake and its basin was shared by Afghanistan, Iran and five countries of the then Soviet Union. The Aral Sea is a modern ecological tragedy; it has lost more than 80 percent of its volume and what is left is salty brine. Canal diversion for irrigation was also (with drought) the major cause of the destruction of Lake Chad – once the sixth largest lake in the world (and the third largest in Africa), now all but gone.

Desalination

The third technology being touted either enthusiastically (by the water industry) or reluctantly (by governments of some water-stressed countries) is desalination. Desalination is the process whereby salt is removed from seawater or brackish water either by evaporation or by forcing the salty water through tiny membrane filters in order to create fresh, drinkable water. According to the International Desalination Association, there are now 12,300 desalination plants worldwide in 155 countries with a collective capacity to produce forty-seven million cubic meters of water a day.

These statistics are not as impressive as they sound. Most desalination plants are small, and used for highly localized, high-valued and industrial needs. Only in a few places, such as the Middle East and the Caribbean, is desalination an integral part of a country's water solution. Two thousand of these plants are in Saudi Arabia, which accounts for one-quarter of the world's desalinated water production. This is not a coincidence; desalination of seawater is very, very expensive and very few water-stressed countries have the resources of that oil-rich nation. Globally, says the Pacific Institute, current desalination plants have the capacity to provide for only three one-thousandths of total world freshwater use.

However, as the global water crisis becomes more evident, many politicians and bureaucrats are looking to this technology for salvation. Some very large plants are under construction in Israel, Singapore and Australia, and there are thirty large-scale ocean desalination plants in the planning stage for California. Global demand is projected to grow by 25 percent every year, according to the International Desalination Association. It is very important, then, to ascertain whether desalination is truly the answer some are claiming.

Any close examination of this technology reveals major environmental and human health hazards. First, desalination plants are highly energy-intensive and put a huge additional burden on local power grids. In *Twenty-Thirst Century*, Australian environmental writer John Archer's scathing book on his country's water crisis, Archer gives the example of a proposed plant in Sydney. Initially it will only be capable of producing one hundred megaliters of water a day – a mere one and a half hours of Sydney's current needs – but will require enough energy to produce 255,500 tons (about 232 million kilograms) of greenhouse gases every year. Worldwide, large-scale desalination technology would radically increase greenhouse gas emissions that, in turn, exacerbate the water shortage crisis the plants were built to alleviate.

Second, all desalination plants generate a lethal by-product – a poisonous combination of concentrated brine mixed with the chemicals and heavy metals used in the production of freshwater to prevent salt erosion and clean and maintain the reverse osmosis membranes. For every liter of desalted water, a liter of poison is pumped back into the sea. Archer notes that the proposed Sydney plant would create more than thirty-six billion liters of waste every year. Aerial photos of the big Saudi Arabian plants show a massive black brine slick fanning out into the ocean, resembling the purple ink discharged from a giant squid. Worldwide, current desalination plants produce twenty billion liters of waste every day. As well, the discharge contains the decomposed remains of aquatic life – such as plankton, eggs, larvae and fish – that are killed during the intake process; these remains reduce the oxygen content of the water near the discharge pipes, creating additional stress on marine life.

Third, the water fed into the desalination system may contain dangerous contaminates that are not filtered out by the reverse-osmosis process. These may include biological contaminants such as viruses and bacteria; chemical contaminants such as endocrine disrupters, pharmaceuticals and personal-care products; and algal toxins such as paralytic shellfish poisoning. These contaminants are found everywhere.

But there is a huge additional problem when desalination plants are built in countries that discharge their waste into the ocean, thereby guaranteeing that much of the intake water will be polluted. Sydney, for instance, dumps a billion liters of sewage into the ocean every day; much of this would be sucked back into the planned desalination plant, where only the salt would be filtered out. This water would then be used for the daily water needs of Sydney's citizens. When we remember that the Third World still discharges 90 percent of its waste, it is not hard to imagine the quality of water that would be processed by desalination plants for human consumption. Desalination plants are also big, bulky plants that block ocean vistas. As well, they are noisy and produce a foul odor.

The Pacific Institute's Peter Gleick is not in principle opposed to desalination technology. Yet in a detailed report on desalination in *The World's Water, 2006–07*, Gleick concludes that, between the environmental concerns and the astronomical costs, this technology is still an "elusive dream" and far less of an answer to the global water crisis than the "soft path" of conservation, reclamation of polluted water, energy efficiency, sustainable agricultural practices and infrastructure investment. John Archer would agree but is more blunt. He writes, "Desalination of the sea is not the answer to our water problems. It is survival technology, a life support system, an admission of the extent of our failure." In a June 2007 review of desalination plants worldwide, the WWF agreed with the Pacific Institute, saying that desalination poses a threat to the environment worldwide and will exacerabate climate change. Large desalination plants may soon become "the new dams," said the WWF, obscuring the need for conservation of rivers and wedlands.

Our Political Leaders Are Failing Us

So here, then, is the answer to the question, Can we run out of freshwater? Yes, there is a fixed amount of water on Earth. Yes, it is still here somewhere. But we humans have depleted, polluted and diverted it to such an extent that we can now actually say the planet is running out of accessible, clean water. *Fast.* The freshwater crisis is easily as great a threat to the Earth and humans as climate change (to which it is deeply linked) but has had very little attention paid to it in comparison.

The world is running out of available, clean freshwater at an exponentially dangerous rate just as the population of the world is set to increase again. It is like a comet poised to hit the Earth. If a comet really did threaten the entire world, it is likely that our politicians would suddenly find that religious and ethnic differences had lost much of their meaning. Political leaders would quickly come together to find a solution to this common threat.

However, with rare exceptions, average people do not know that the world is facing a comet called the global water crisis. And they are not being served by their political leaders, who are in some kind of inexplicable denial. The crisis is not reported enough in the mainstream media, and when it is, it is usually reported as a regional or local problem, not an international one. Water policy is raised as a major issue in very few national elections, even in water-stressed countries. In fact, in many countries, denial is the political response to the global water crisis.

In November 2006, former Australian prime minister John Howard hosted a high-level summit in Sydney to deal with what one scientist called "the worst drought in Australia in 1,000 years." Howard's answer? Allow farmers to "trade" country water to the city, thereby draining already thirsty rivers of yet more water; drain the wetlands to supply the cities; ship in tankers full of water from Tasmania; and look to technology such as desalination plants. The government uttered not a word about conservation, protecting watersheds and replenishing water systems, cleaning up toxic dumps or stopping the massive export of Australia's water stock-in-trade with China.

Under two terms of the Bush administration, environmental stewardship has been dealt a terrible blow. In his passionate book *Crimes Against Nature*, Robert F. Kennedy Jr. reports that the Bush White House has rolled back more than four hundred pieces of environmental legislation and taken the United States back to a time before environmental consciousness. Not only has George W. Bush not taken his country's water crisis seriously, he has cut funding for clean water and safe drinking programs and allowed once-banned chemicals and toxins back into circulation, gutting the Clean Water Act. He has allowed logging and mining in national parks, resulting in the destruction of pristine rivers and lakes. Funding for water research in the United States has been stagnant for thirty years, and the portion dedicated to water quality has actually been reduced in the last decade.

Canada has no national water act and no inventory of its groundwater resources. A 2005 report from Environment Canada said that a national water crisis was looming and that no one in government seemed to be listening. The report gave a blunt assessment of pollution and overextraction of Canada's water systems and noted a total lack of leadership on the issue by both federal and provincial governments. Canada is allowing the destruction of huge amounts of water in the Alberta tar sands, where water is actually being lost to the hydrologic cycle in order to mine the heavy oil from the ground.

To its credit, Europe has taken some more serious action. In 2000, the European Commission launched the Water Framework Initiative, a European Union–wide plan for water conservation, clean up and administration based on the joint management of river basins. All European waters must achieve "Good Status" by 2015.

All people in the European region must have access to clean drinking water (there are currently 120 million without), and the environment must be protected as well. The initiative requires cross-border cooperation on all areas of watershed protection. While this program is among the most progressive in the world, the powerful countries of Europe have been responsible for practices in the Third World that have denied clean water to millions (see Chapter 2). Europe's record must include this fuller picture.

In the developing world, all that most governments can do is desperately try to provide water for their citizens. There is little attempt to address the environmental crisis that has polluted water in the first place. Most have bought into the tenets of the World Bank and the World Trade Organization and are attempting to export their way to prosperity, creating more environmental damage in the process. And most are helpless to police the big transnational oil, forestry and mining corporations fouling their water systems; some are in collusion with these companies to repress their own people. Most First World governments refuse to even consider legislation that would hold their corporations accountable for polluting the water systems of poor countries.

The United Nations, the European Union and the World Bank have devised a water rescue plan for the developing world totally devoid of plans to deal with the growing rivers of sewage killing whole watersheds and coastlines. Ninety percent of the raw sewage in poor (and some not so poor) countries is still discharged untreated. Most of the megacities in the Third World also lose massive amounts of water from leaky infrastructure. In the global South, more than 50 percent of municipal water is lost because of faulty systems.

Nor are most rich countries prepared to cancel or at least renegotiate the debt owed by the global South to the global North to allow governments in poor countries to address these issues themselves. Every year, more money flows to the global North to pay the debt than flows to the global South in aid and trade together. No serious plan to alleviate the water crisis can ignore the poverty of the global South and the role of debt repayment in that cycle.

In addition, few countries in the world are confronting the pervasive and harmful agricultural practices that are dramatically exacerbating the crisis. Large-scale factory farms create a staggering amount of manure and depend on intensive use of antibiotics, nitrogen fertilizers and pesticides, all of which eventually end up in the water supply. Flood irrigation, used in many parts of the world, wastes enormous amounts of water. (In China, close to 80 percent of water used in flood irrigation – the main form of irrigation in that country – is lost to evaporation.) Flood irrigation also leads to desertification, as it overtills the soil, which then is carried away by the wind. Yet not only are wealthy countries wedded to industrial agriculture, the World Bank and the World Trade Organization promote this model in the developing world.

Nor have these international institutions and the powerful countries behind them, blinded by their unquestioning faith in market economics, begun seriously to question the abuse and overuse of water by industry. While it is commonly understood that agriculture is the biggest user of water in the world, this is changing. In industrialized countries, industry now accounts for 59 percent of total water withdrawals, and industry is fast gaining as a water abuser in developing countries as well. India, for instance, will triple its use of water for industry in the next decade. As countries such as China, India, Malaysia and Brazil undergo industrialization at an unprecedented rate, water use and misuse is growing exponentially. Yet few political leaders have the courage or foresight to question this model of development.

～～～

Every day, the failure of our political leaders to address the global water crisis becomes more evident. Every day, the need for a comprehensive water crisis plan becomes more urgent. If ever there was a moment for all governments and international institutions to come together to find a collective solution to this emergency, now is that moment. If ever there was a time for a plan of conservation and water justice to deal with the twin water crises of scarcity and inequity, now is that time. The world does not lack the knowledge about how to build a water-secure future; it lacks the political will.

But not only are our political leaders following the false promises of a quick technological fix, they are abdicating the real decision-making about the future of the world's depleting water supplies to a group of private interests and transnational corporations that view the crisis as an opportunity to make money and gain power. As we'll see, these big players know where the water is. They simply follow the money.

Chapter 2

Setting the Stage for Corporate Control of Water

> *"Hey, Maasai, do you think privatization is going*
> *to change the economic situation of this country?"*
> *"Yes! We still have our country and it is at a stand-*
> *still. – If there's no capital we will die of starvation! –*
> *And there are people with capital in the world and these*
> *are plenty! – It's better that we call them and benefit."*
> *Chorus: "We need money . . ."*
>
> – Ebbo, Maasai warrior and Tanzanian rapper,
> co-written and produced by the
> Adam Smith Institute and the World Bank

While full knowledge of the extent of the global water crisis and its documentation are recent – and are only now beginning to permeate mass consciousness – one sector has had its eye on the world's dwindling water resources for decades and has been quietly extending its reach into every aspect of water. What the private sector understands is that in a world running out of clean water, whoever controls it will be both powerful and wealthy.

Water, of course, has traditionally been viewed as a public resource. Increasingly, however, freshwater supplies are being privatized in a whole range of ways as a powerful water industry moves to create a cartel resembling the one that now controls every facet of energy, from exploration and production to distribution.

Private, for-profit water companies now provide municipal water services in many parts of the world; put massive amounts of freshwater in bottles for sale; control vast quantities of water used in industrial farming, mining, energy production, computers, cars and other water-intensive industries; own and operate many of the dams, pipelines, nanotechnology, water purification systems and desalination plants governments are looking to for the technological panacea to water shortages; provide infra-structure technologies to replace old municipal water systems; control the virtual water trade; buy up groundwater rights and whole watersheds in order to own large quantities of water stock; and trade in shares in an industry set to increase its profits dramatically in the coming years.

All of these developments are fairly recent. Thirty years ago, only a small elite drank bottled "mineral" water. Water technologies were in their infancy. Water pipelines for large water diversion were almost nonexistent. The vast majority of the industrialized world's water services were (and still are) delivered by state-owned public utilities while huge numbers of people in the global South still lived in rural communities, relying on local rivers, lakes and wells for their water. No one could imagine a time when water would cost more than gasoline or would be traded like shares on the stock exchange.

In the global North, public delivery of water helped to create the political stability and financial equity necessary for the great advances of the industrial age. During the late nineteenth and early twentieth centuries, the industrializing countries of Europe and North America, as well as Australia (and later Japan), adopted universal public water and sanitation services to protect public health and promote national economic development. Public systems allowed municipalities to take out long-term loans at better rates than were available to private contractors, which, in turn, allowed them to extend water services as their communities grew. With a few exceptions, these countries still provide public water systems of which they are very proud.

France was the notable exception. Since the late 1800s, it had encouraged the creation of a private water industry whose principle operators – Lyonnaise des Eaux, later to become Suez, and General des Eaux, later to become Vivendi, then Veolia – were perfectly poised to take advantage of the push for water privatization and would soon become the most powerful water transnational corporations in the world. But as Public Services International points out, even in France, the cost of building and extending water and sanitation networks was met through public financing.

The story in the global South was quite different. Unlike in the global North, services lagged in Africa, Asia and Latin America, where a colonial legacy had created urban water services only for the elite. As a consequence, many millions of poor urban dwellers had no access to water or sanitation, which led to terrible outbreaks of disease. This was exacerbated in the last thirty years with the exodus from rural communities to the burgeoning mega-cities of the Third World. This exodus, combined with growing pollution of surface waters, created new demands for water serv-ices, demands that could not be met by governments crippled by growing poverty and increasing debt.

By the early 1980s, it was clear that a crisis of major propor-tions was emerging. In response, the UN declared the 1980s to be the International Drinking Water Supply and Sanitation Decade and set targets for the provision of water to the South, originally based on the public Northern model. By the end of that decade, however, a public model for the developing world had been aban-doned in favor of a private model that, not coincidentally, would benefit the private water companies of Europe. This was not a random development. A private model for the global South was planned and carried out by some of the most powerful forces in the world.

Water Privatization Is Forced on the Global South

The shift from a public to a private model of water services can be traced to the rise of a neo-liberal, market-based ideology first manifested in Margaret Thatcher's Britain, then adopted by Ronald Reagan in the United States as a major component of his war against communism. By the late 1970s, the stage was set for an emerging global regime based on the belief that liberal market economics constitutes the one and only choice for the whole world, including the developing world. Northern governments began to give up foreign investment controls, liberalize trade, deregulate their internal economies, privatize state services and utilities and enter into head-to-head competition. Soon, this Washington Consensus became the guiding mantra for the elite running the global institutions involved in water development, including the World Bank, the International Monetary Fund and even the United Nations.

In 1989, Thatcher privatized Great Britain's publicly owned regional water authorities, which were sold off to private companies at bargain prices. As Ann-Christin Holland explains in her book *The Water Business*, these sales included large properties with significant cultural and natural assets. In fact, the private companies became owners of the entire infrastructure, including the buildings. They were given licenses to run the water systems without competition for twenty-five years as well as free rein to charge what they liked, lay off employees and make as much profit as they could.

Thousands of workers were laid off, water rates were jacked up and pre-tax profits soared by 147 percent in the first decade of privatization. Millions had their water cut off when they couldn't pay their water bills, a practice Tony Blair ended when he came to power in 1997. (In January 2007, however, the British government announced it would introduce compulsory water metering in water-stressed areas of the country. The new rules will affect nineteen million people.)

In spite of the obvious failures of water privatization in England, it is this model – not the far more successful model of public delivery entrenched in most of the global North – that was exported to the developing countries of the global South. In taking the lead in water privatization, Thatcher also helped create several other private water companies that would be poised, along with Suez and Veolia, to jump into the international private market. Most notable was Thames Water, which was bought out by the German energy giant RWE in 2002 to become RWE Thames, the third largest water corporation in the world.

Prior to this, during the 1980s, the World Bank began to abandon its policy of national development in the South in favor of a new policy of development designed to force poor countries to adopt the economic model of the Washington Consensus. Most had been loaned money at low interest rates but found themselves unable to meet debt repayment schedules when interest rates soared. The World Bank agreed to renegotiate their loans on condition that the countries in question undergo Structural Adjustment Programs that required them to sell off public enterprises and utilities and privatize essential public services such as healthcare, education, electricity and transportation. (In spite of monumental sacrifices, Third World debt has grown by 400 percent since 1980.)

It was only a matter of time before water and sanitation services were targeted for privatization. By the early 1990s, the World Bank, the International Monetary Fund and the other regional development banks, including the Asian Development Bank, the African Development Bank and the Inter-American Development Bank, were encouraging poor countries to let the big European water corporations run their water systems for profit. A country's ability to choose between public or private water systems was steadily eroded, and by 2006, the vast majority of loans for water were conditional on privatization. In fifteen years, reports Public Services International, there was an 800 percent increase in African, Asian and Latin American water users purchasing water from transnational water companies.

Inside the World Bank

First World countries control the World Bank and have voting power proportional to the amount they invest in it. Accordingly, the United States (followed by Japan, Germany, the United Kingdom and France) dominates the decisions about who receives the approximately US$20 billion a year loaned out to poor countries and the conditions they must agree to meet in order to receive that money. Funds earmarked for water and sanitation amount to US$3 billion a year. The World Bank uses its power to open markets in the global South for northern corporations. A World Bank Article of Agreement actually states that a principal goal is the promotion of private investments. (A top U.S. Treasury official once famously bragged to Congress that for every dollar the United States contributes to the World Bank, American corporations receive $1.30 back in contracts.)

Although it had been promoting water privatization as one option several years prior to 1993, in that year, the World Bank adopted the *Water Resources Management* policy paper, which noted the "unwillingness" of the poor to pay for water services and stated that water should be treated as an economic commodity, with an emphasis on efficiency, financial discipline and full-cost recovery. (This is a principle that says that corporations can set water prices high enough to not only recover the cost of their investment, but to make profits for their investors.) Increasingly, loans for public projects were rejected in favor of a private model; between 1990 and 2006, the World Bank funded more than three hundred private water projects in the developing world.

There are three basic types of water utility privatization. *Concession* contracts give a private company a license to run the water system and charge customers to make a profit. The private company is responsible for all investments, including building new pipes and sewers to connect households. The British model of privatization is a type of concession (although in this case, the complete system was sold off through public shares). India practices a form of extreme concession whereby whole river systems are leased to the companies who run them for profit without government interference. *Leases* are contracts under which the company is responsible for running the distribution system and for making the investments necessary to repair and renew the existing assets, but the local government remains responsible for new investment. *Management* contracts make the private company responsible only for managing the water service but not for any investments.

As the World Development Movement points out, the World Bank uses the term *privatization* only when referring to the complete divestiture of public assets, preferring the less politically loaded terms of *private sector participation* or *public-private partnerships* to describe its more current projects, most of which are leases or management contracts. The notion of a partnership has the ring of democracy and shared responsibility. But these contracts should all be considered privatizations because they all involve profits for the private companies and cutoffs to people who cannot pay for their "product." As well, the government and community "partners" – that is, the people who live in the communities in question – have no alternatives if the so-called partnership fails. But the corporate partner will (and regularly does) leave the partnership if the profits dry up.

The World Bank promotes private water services in the global South through several of its component agencies: the International Bank for Reconstruction and the International Development Association, which lend money to poor countries (and advantageous loans to the poorest) based on conditions that the countries adopt a private water delivery model; the International Finance Corporation and the Multilateral Investment Guarantee Agency, which encourage private investors to invest in the water sector in poor countries, and, in the case of the latter, actually insures those investors against risks of all kinds, including local political resistance; and the International Centre for Settlement of Investment Disputes (ICSID), an arbitration court used by water companies to sue governments who try to break their contracts. (According to an April 2007 Food and Water Watch report, *Challenging Corporate Investor Rule*, nearly 70 percent of ICSID cases are settled in favor of the investor, with compensation awarded against the country where the investment failed. In at least seven cases, the investors' revenues exceeded the gross domestic product of the country they were challenging.)

Countries are encouraged through these various mechanisms to adopt a private water services model, with a "carrot and stick" approach (the carrot being both debt relief and the funds themselves; the unspoken threat of the withdrawal of aid being the stick). In many cases, the agreements made among the World Bank, the water corporation and the country in question are completely secret and the terms of the deal are not accessible to citizens. Increasingly, throughout the 1990s, the emphasis was on full-cost recovery for the companies. By 2003, reported Public Citizen, 99 percent of loans promoted full-cost recovery for the private companies.

Water privatization also became a key component of the World Bank's Poverty Reduction Strategy Papers (PRSPs), the primary strategic and implementation vehicle used to reach the UN Millennium Development Goals and the framework agreements through which developing countries receive international aid. Poor countries must complete a PRSP to receive debt relief through the Heavily Indebted Poor Country Initiative, which usually takes the form of agreeing to adopt neo-liberal market reforms and promising not to use the aid money for poverty reduction or public services such as healthcare, education or water delivery. Through PRSPs, countries agree instead to promote economic growth through macroeconomic policies and infusions of foreign direct investment as well as the selloff of state-owned enterprises and utilities. The World Development Movement studied the fifty PRSPs signed off by the World Bank in the first half of 2005 and found that 90 percent of the countries promised more privatization in general and 62 percent promised water privatization specifically.

(Regional development banks such as the Inter-American Development Bank, the Asian Development Bank and the African Development Bank follow the same kinds of policies and promote water privatization in much the same way as the World Bank.)

The World Bank Manufactures
Global Consensus on Privatization

How the World Bank and other global financial institutions came to impose this new model of water delivery on the global South is an important story. It did not go unnoticed in poor countries that most of the northern countries behind the World Bank still held on to cherished public water services and had no intention of giving them up. As well, most poor countries had already had terrible experiences with structural adjustment policies and the forced abandonment of public health and education programs, for instance. To successfully sell wholesale privatization to a population suffering from terrible water shortages required a highly orchestrated plan that would intimately involve the elite of the targeted countries.

Sociologist Michael Goldman of the University of Minnesota has analyzed how the World Bank and the big water companies set out to promote a major shift in water policy over a relatively short time, actively seeking the buy-in of non-governmental organizations (NGOs), think tanks, state agencies, the media and the private sector across the global North and South. Through its Water Policy Capacity Building Program, the World Bank Institute (the "capacity building" arm of the bank that promotes bank values and programs through education and outreach) has put thousands of parliamentarians, policymakers, technical specialists, journalists, teachers, students, civil society leaders and Third World elites through intensive programs on private water management; these "experts" then returned home to promote a private model of water delivery to their governments. (It is important to note that economic globalization has created a "First World" class in the developing world, as well as a "Third World" class in the developed world, one that often has more in common with its class in other parts of the world than with its fellow citizens.)

A great deal of work, money and planning has gone into acquiring the "manufactured consent" of the global ruling class to water privatization. Quite simply, says Goldman, since the mid-1990s, in the name of poverty alleviation, water privatization has become a key green neo-liberal project for the World Bank, which has cultivated "elite transnational water policy networks" to create the appearance of a worldwide consensus on a private future for water. The uniquely situated and well-funded members of this elite network have filled every policy space, for who else, asks Goldman, can afford to attend expensive global forums, speak up with reliable global data and sit on the powerful round-tables on water?

Their consensus: debt and poverty are not the problem. The main problem with degraded water services in the Third World is inefficient and corrupt governments whose failure to protect water, so as to reflect its true cost, has led to a culture of waste-fulness among the masses. The poor lack access to water because of their irresponsible governments, goes the refrain; the World Bank and its private sector colleagues are simply on an ethical mission of poverty alleviation, ecological sustainability and social justice. In fact, these projects were presented as a bailout by foreign firms willing to help indebted and floundering public agencies meet World Bank targets – corporations fulfilling the role of charitable trusts offering a helping hand, technology trans-fers and expertise. (This altruistic tone changed, however, by 2003, due to massive local resistance to their presence in commu-nities all over the world. Now the big water companies are telling the banks that if they are not underwritten by guaranteed global financing, they will abandon large parts of the global South for greener markets.)

Elites in government, private business and academia in the global North regularly assert that northern-style capitalism will do for the global South what their governments, mired in arrested development and corruption, cannot. Cloaked in this mantle of self-righteousness, the World Bank can refuse, with a clear con-science, the demands of poor countries when they seek funding for public, not private, water services. Where persuasion fails, there is the stick of conditionality: accept one of the major water corpora-tions or go without funding.

United Nations

To be truly successful in promoting water privatization in the global South, the World Bank and the big water companies had to enlist the support of the United Nations and set up formal global institutions to promote their agenda. At an important UN confer-ence held in Dublin in January 1992 and attended by government officials and NGOs from one hundred countries, it was declared that water has an "economic value" in all its "competing uses" and should be recognized as an "economic good." Water was being wasted because people didn't have to pay for water, conference participants agreed, and therefore some form of user fee was necessary to curb this waste. No official recognition was given to the fact that in the global North, water waste is rampant compared to the global South. Yet both the criticism and the new pricing rules were clearly aimed at the global South. This was the first time that water had been described as an economic good in any UN forum or publication. But it would not be the last.

Since the Dublin meeting, the United Nations, under former secretary-general Kofi Annan, has promoted private sector involvement in water services in a number of ways. Both Suez and Veolia are charter members of the United Nations Global Compact, an initiative to encourage corporations to adopt voluntary human rights and environmental standards. The compact has been widely criticized as "blue washing" for giving UN approval to companies with public relations problems for serious violations in these areas, such as Shell Oil and Nike. At the July 2000 launch of the compact, Annan recommitted the UN to support "free trade and open global markets" and admitted to reporters that the UN has no way of enforcing any standards.

Veolia and Suez funded a UNESCO conference in October 2002 concerning legal frameworks on water, which resulted in a report carrying the logo of the UN and both water corporations. That same year, Suez gave US$400,000 to UNESCO's water research institute, located at the Delft University of Technology, Netherlands, in part to pay for financing a professional chair on the subject of public-private partnerships. The money gave Suez direct influence over the design of the curriculum and a high level of involvement in the masters of water management program taught at the institute. Suez also helps fund the UNESCO Chair for Integrated Water Resource Management in Casablanca, Morocco. As well, Gerard Payen, former CEO of Suez's water division, is a current member of the UN Advisory Board on Water and Sanitation. So it can come as no surprise that the UN Millennium Development Goals (MDGs), set at a September 2000 meeting of the General Assembly, were flawed from the beginning because of the deep involvement of the water transnationals. The freshwater component of the MDGs – namely, to cut by half the proportion of people living without safe drinking water and sanitation by 2015 – is now further away than ever.

World Trade Organization

The World Trade Organization (WTO) was created in 1995 to administer a number of international trade agreements involving goods, food, patents, intellectual property rights and services, and enforces an extensive body of rules intended to limit the power of governments and increase opportunities for transnational business trade. Under the rules of one agreement – the General Agreement on Tariffs and Trade (GATT) – water is included as a "good" and, as such, is subject to the rule that prohibits the use of export controls for any purpose and eliminates quantitative restrictions on imports and exports. In practical terms, this means that, once a country has launched commercial water exports, it cannot change its mind based on environmental concerns and restrict the flow of water out of its territory. This will be very helpful to the industries in the water export and pipeline-building business.

As well, the WTO has embarked on an ambitious new agreement called the General Agreement on Trade in Services (GATS), whose explicit intention is to liberalize all service sectors in all the WTO member countries to allow for private competition in sectors once controlled exclusively by governments. Dozens of types of water services are already included in the GATS, including environmental services, wastewater treatment, purification systems, construction of water pipes, groundwater assessment, irrigation and water transport services, to name a few. Governments can no longer maintain these areas under public sector control or even favor non-profit delivery of these services. Corporations operating in these sectors are also aided by the WTO's Smart Regulation process to create one set of global regulatory standards for all trade transactions. The WTO says that Smart stands for Specific, Measurable, Attainable, Realistic and Timely but is really a corporate-led initiative to create a "level playing field" for business with the fewest regulatory barriers and the lowest set of common standards.

A recent proposal of the WTO was to add drinking water to the GATS, which would mean that any municipality in the world that decided to try a private system for water delivery would not be allowed to change its mind (as many have done) and revert to a public water system without the unanimous consent of all the other 150 countries of the organization.

World Business Council for Sustainable Development

The World Bank also needed some powerful business allies aside from the water companies. The World Business Council for Sustainable Development (WBCSD), a corporate lobby network of 180 corporations as well as more than fifty national and regional business councils, has become a key player in the transnational water policy network. It was formed in 1992 to influence the outcome of that year's Rio Earth Summit and, true to its mandate of opposing attempts to place international rules on global business transactions, was largely credited with watering down many of the resolutions coming out of that gathering. Along with the International Chamber of Commerce, the WBCSD succeeded in having references to mandatory environmental regulations entirely eliminated from the summit's Agenda 21 document, placing the emphasis instead on corporate "self-regulation."

In 1997, the WBCSD set up a formal "water working group," bringing together corporations from mining and metals, oil and gas, food and beverage, finance and equipment, as well as, of course, water services to influence global policy on water. This group played a major and destructive role at the 2002 World Summit on Sustainable Development, where it released a report, *Water for the Poor*, which called for the accelerated privatization of services and full-cost recovery to private companies for providing water. The report openly states, "Providing water services to the poor presents a business opportunity. New pipes, pumps, measurements and monitoring devices, and billing and record keeping systems will be required to modernize and expand water infrastructure . . . this program has the possibility of creating huge employment and sales opportunities for large and small businesses alike." In 2006, the WBCSD published a series of future scenarios called *Business in the World of Water*, which challenged businesses on their "global fitness in the marketplace" in a water-deprived world.

Public-Private Infrastructure Advisory Facility, Water and Sanitation Program and USAID

The World Bank also needed the support of the international development agencies of rich countries in order to direct its overseas aid on water development to private water models. In 1999, together with the World Bank, the United Kingdom's Department for International Development set up the Public-Private Infrastructure Advisory Facility (PPIAF) to promote private sector involvement in the use of aid money targeted for water services and to provide consultants to developing countries to build consensus for water "reform" inside their governments and with the public. In an early memo to the British government, the new agency explained that its work would mean that governments would have to change their role as well, no longer directly providing water services but "mastering the new business of fostering competition among private providers, regulating where competition is weak and supporting the private sector generally." Soon Japan, Canada, France, Germany, Italy, Netherlands, Norway, Sweden, Switzerland, the United States and the Asian Development Bank joined the United Kingdom in this project, which is organized and administered directly by the World Bank. Austria and Australia are both close to joining.

In a November 2006 report, *Down the Drain: How Aid for Water Sector Reform Could Be Better Spent*, two major NGOs, the Association for International Water and Forest Studies (FIVAS) of Norway and the World Development Movement of Great Britain, outline the clear role played by this agency over the last seventeen years. PPIAF funds consultants to advise poor governments on the changes required to domestic legislation, policy institutions and regulations so as to attract private water companies and secure the support of the World Bank. Sometimes, consultants are brought in to rescue a botched private sector operation. Often, they are brought in to build consensus on the need to privatize water services. In early 2000, for example, in response to growing criticism of water privatization schemes, PPIAF funded a program attended by journalists from nine African countries to "increase press coverage related to water issues in Africa and to improve the quality and objectivity of this coverage."

FIVAS and the World Development Movement point out that the work of this agency – which has funded pro-privatization projects in thirty-seven poor countries at a cost of close to US$19 million – is flawed because it lacks transparency, is ideologically based, rejects public sector options, quashes legitimate dissent and imposes the interests of northern consulting firms and water corporations on the global South.

Another World Bank/United Nations/northern development agency donor initiative is the Water and Sanitation Program (WSP), which has "evolved" from a United Nations Development Program agency that provided low-tech water service devices such as hand pumps and latrines to a significant partner with the World Bank in working with the private sector to "effect the regulatory and structural changes needed for broad reform" in the global South. The WSP got a huge public relations boost in 2006 when the Bill & Melinda Gates Foundation made a US$30 million grant to the fund. The Water and Sanitation Program supports the Water Utility Partnership, which promotes public-private water partnerships in Africa, and hosted a series of government seminars to encourage African countries to adopt "greater private sector involvement in the water sector" as a precondition of funding.

PPIAF is joined by its American counterpart, the U.S. Agency for International Development (USAID), which uses its aid dollars to advance U.S. foreign policy in developing countries and whose key objective is to expand democracy through free markets. USAID openly supports private water delivery in the global South and is behind many private projects in India, Latin America and Africa. In March 2002, reports Public Services International, a number of private water companies launched a new African organization called Partners in Africa for Water and Sanitation specifically to promote water and sanitation services in South Africa, Nigeria and Uganda. The new group used a report by an American consultancy firm, PADCO, to convince municipal politicians in these countries that privatization was the best option for them. The report was financed by USAID.

Global Water Partnership
In 1996, two very powerful new global water institutions were created to consolidate the new corporate water model and create a new space where all of the members of the transnational water policy network could work together. Italian water expert and activist Riccardo Petrella calls them the new "global high command of water," so powerful have they become in dictating global water policy. The Global Water Partnership was formed by the World Bank, the United Nations Development Program and the Swedish International Development Cooperation Agency and operates as a clearing house and alliance-building instrument among governments, the private sector and civil society to promote global water management based on the Dublin principles.

The World Bank, the United Nations and the international development agencies of many northern countries fund this organization, which now has branches all over the world. The Global Water Partnership was instrumental in launching the controversial 2003 report *Financing Water for All*, which recommended using public funds to guarantee profits to private water companies operating in areas where they were meeting local resistance (see below).

World Water Council

The World Water Council (wwc) calls itself an "international water policy think tank" but it is far, far more than that. Sponsored by the World Bank and the United Nations, the wwc uses its power and prestige to promote private water delivery to governments around the world. The private water sector and other private corporations dominate its three-hundred-plus membership list. Amid a sprinkling of government agencies and NGOs are corporations and industry associations from many sectors: private water operators, engineering, construction, hydropower, dams, irrigation, infrastructure and wastewater treatment, desalination, investment banks and public affairs consultants. All of the big transnational water corporations are members, as is the International Water Association with 400 corporate members in 130 countries. Even Price Waterhouse Coopers, the global management-consulting giant with 150,000 staff operating in 152 countries, is a charter member of this powerful organization, having recently entered the lucrative business of water.

The reason: with the blessing of the United Nations and the development agencies of the industrialized countries, the private sector can advance its corporate interests through the World Water Council in the name of poverty alleviation and sustainable development – among the stated goals of the council. The reality is that the World Water Council has become, along with the Global Water Partnership, a major vehicle for the corporate takeover of the world's water. Its president is Loïc Fauchon, president of Groupe des Eaux de Marseille, owned by Suez and Veolia, and its vice-president is René Coulomb, another senior director at Suez. The World Water Council held its founding meeting in Marrakech in March 2000 and has sponsored a giant forum every three years since: The Hague in March 2000, Kyoto in March 2003 and Mexico City in March 2006, all heavily attended by the water corporations. The next forum is slated for Istanbul in March 2009.

AquaFed

A recent player in the elite transnational water policy network is AquaFed, the International Federation of Private Water Operators, a new lobby group started by the big European water utilities. It was formed in October 2005 "to connect international organizations" such as the United Nations, the World Bank and the European Union "with private sector providers of water and wastewater services." AquaFed now has more than two hundred water and wastewater service company members from thirty countries, including Suez, Veolia and United Water, as well as several national associations of water operators, including the U.S.–based Water Partnership Council. AquaFed's president is Gerard Payen, former chairman of Suez's water division. Jack Moss, senior water adviser at Suez, represents AquaFed at international meetings.

AquaFed claims that "until now, private water operators as a body have not been represented at an international level" – an odd claim, given the prominence of these corporations in the World Water Council and the World Business Council for Sustainable Development, to name just two other networks. However, as David Hall of the Public Services International Research Unit and Olivier Hoedeman of Corporate Europe Observatory point out, water privatization has come under intense scrutiny and criticism in recent years, and the corporations are likely seeking a more direct way to push their agenda than the World Water Council, which also has representatives from government. AquaFed has two offices – one right across from the European Union headquarters in Brussels and the other in the heart of Paris. The choice of office venue is not accidental; this lobby group wants to deepen the already close ties between European Union politicians and bureaucrats, who have come under growing pressure to abandon their pro-privatization position, and the industry.

Non-Governmental Organizations

The final sector in this elite network includes several prominent environmental non-governmental organizations that work within environmental non-governmental organizations that work within the established global institutions, including the World Bank and the World Water Council. These include the highly influential, London-based WaterAid founded by the British water companies, which provides water services in Africa and Asia; Freshwater Action Network (FAN), a global network of environmental and community groups now exploring "dialogue" between civil society and the World Bank; the World Wildlife Fund, one of the world's largest conservation groups; and Green Cross International, an environmental and education organization led by Mikhail Gorbachev, which works with the World Water Council to promote a UN convention on the right to water that would endorse private financing for water projects.

There is an ongoing dialogue in the NGO community about whether to work within these global financial institutions. Many in civil society, particularly the grassroots groups in the global South fighting the big water companies, consider any dialogue with the World Bank or the World Water Council to be a waste of time at best and a sellout at worst. NGOs working within the system explain their working relationship with the World Bank and the World Water Council as a practical way to influence their policies and thinking, while at the same time working toward the same goals as civil society groups who agitate from outside these spheres of influence. It should be noted that a number of these organizations, particularly WaterAid, are increasingly critical of the World Bank and its pro-privatization policies.

Water Policy Consensus Is Shaped at Giant Global Forums

Since 2000, the World Water Council, the United Nations and the World Bank have hosted a series of high-profile international summits on water, which purport to be neutral in their ideological perspective and open to all "stakeholders," but which are really designed to ensure consensus on the benefits of privatization. Importantly, each of these summits hosted a Ministerial countries attending to assess global opinion on water policy and take that assessment back to national legislatures.

Second World Water Forum – The Hague, March 2000

Close to 6,000 people from all over the world attended the World Water Forum in The Hague in March 2000, as well as 500 journalists and government representatives from 130 countries. Attracted by topics ranging from gender equity in water access to preserving watershed integrity, thousands of local community representatives believed they were going to be part of a real dialogue on the world water crisis. Instead, they found a tightly controlled agenda with panels held in huge auditoriums extolling public-private partnerships and speakers from only the major water corporations, including Suez and Vivendi (Veolia's predecessor), the major bottled water companies, including Nestlé, and the World Bank. No civil society organizations were given a role on stage or off, and all dissent was resisted.

It was in The Hague that the World Commission on Water for the 21st Century, set up two years previous by the World Water Council and the World Bank, issued its very public *World Water Vision: A Water Secure World*, which made the now utterly discredited prediction of a 620 percent increase in private sector water investments over a thirty-year period, a sum that would outshine public investment three to one. Given the makeup of the commissioners, who included Ismail Serageldin of the World Bank, Enrique V. Iglesias of the Inter-American Development Bank and Jerome Monod, chairman of Suez, it came as no surprise that the "vision" concluded that "consumers" in the Third World had to start paying for water and that, where governments are unable to pay for the needed infrastructure, the private sector should be encouraged. Further, the World Commission recommended – for the first time in an official UN-related document – full-cost pricing for water services, meaning that consumers would have to pay not only for the cost of their water, but enough that the "investors" would recover a profit.

The World Water Forum in its final declaration refused to recognize water as a human right, instead insisting that it is a "human need," just as easily delivered by private companies as by governments.

World Summit on Sustainable Development – Johannesburg, August 2002

The World Water Council and the World Business Council for Sustainable Development were major players at the biggest water summit of them all – the United Nation's World Summit on Sustainable Development – "Rio + 10" – held in Johannesburg, South Africa, in the late summer of 2002. (Although many issues, such as food security, poverty and the environment were on the agenda, it was water and sanitation – and their attendant profit-making opportunities – that dominated this summit.) Sixty-five thousand delegates and observers from governments, international institutions, NGOs and business as well as thousands of journalists came together to assess the success or failure of the original Earth Summit and write the blueprint for the work to come. The WSSD, however, became totally captive to the interests of transnational corporations and is generally acknowledged by all but the big business community to have been a complete failure.

That this was a show put on by and for corporations was evident as soon as delegates arrived at the airport to see a gigantic De Beers "Water Is Forever" billboard, an obvious play on its "Diamonds Are Forever" ad campaign. De Beers was one of the official corporate sponsors of the US$75 million summit, as were Coca-Cola, McDonald's and BMW. More than one hundred CEOs and seven hundred business delegates from more than two hundred major corporations attended the summit, flooding delegates with glossy brochures touting their newfound "corporate responsibility" ethic. The gathering was held at Sandton, the most exclusive suburb in all of Africa and its financial heart, with gleaming office towers, five-star hotels and a glittery nightlife of posh bars and restaurants. Across a small river (covered in cholera-warning signs) from Sandton sits the township of Alexandra, one of the poorest slums in Africa, where children sift through garbage for food and line up at filthy pipes for water.

To get to the convention site, delegates had to cross through a massive shopping mall with a square in the middle where BMW was showcasing a huge "sustainability bubble" on hydrogen power. A newswire story reported that VIP delegate guests at just one of the many five-star hotels had access to 80,000 bottles of water, 5,000 oysters, more than 373 kilograms of lobster, 820 kilograms of prime beef, 820 kilograms of chicken breasts, 165 kilograms of salmon, 410 kilograms of bacon and sausages and 82 kilograms of a tasty South African fish called kingclip. The US$1,100-a-night price for a suite in this hotel is ten times the average monthly wage in Johannesburg.

All the big water companies were prominent at the WSSD, both as members of the official European government delegations and at a giant trade show called the WaterDome they sponsored to advertise their operations and promote new opportunities for business. (At the gala opening of the WaterDome, hosted by Nelson Mandela and the Prince of Orange, young people dressed as water drops – or tear drops – flitted from corporate booth to corporate booth to provide the requisite "cultural" element.) The water companies wanted to cash in on the lucrative contracts that would open up if the summit endorsed private-public partnerships as the main delivery model to implement the UN Millennium Development Goals (MDGs) and they wanted this access sanctioned by the United Nations and the 189 governments present at the Johannesburg gathering.

In this quest, they were supported by a major new US$1.9 billion undertaking of the European Union called the EU Water Initiative, whose stated aim was to create positive conditions for the private sector in the implementation of the MDGs and that was publicly launched with much fanfare at the WSSD. In all, and with the blessing of the United Nations, 220 partnerships among big business, the World Bank, the International Monetary

Fund and developing countries were announced at the WSSD, the majority for the delivery of water and sanitation services to the global South.

Corporate Europe Observatory, a respected European research think tank, noted in a post-summit analysis that the UN leadership did its best under a hail of criticism to pretend that the non-binding goals it reached meant something real, but that, in the end, the summit failed miserably to make any progress on the urgent social and environmental issues of our time. The only concrete outcome of Johannesburg was that governments and the UN cemented their relationship with the corporate elite and paved the way for years of partnership hype and "greenwash."

Third World Water Forum – Kyoto, March 2003

Seven months later, 24,000 participants, 1,000 journalists and 130 government ministers converged on Kyoto for the Third World Water Forum, where the refrain from Johannesburg was clear: a global consensus has emerged that the private sector is best placed to run the water systems of the global South. Although the organizers of the Kyoto forum were more open to civil society, and some critics were even given a spot on the official program, the pressure for delegates to join the pro-corporate refrain was just as strong as in The Hague. Once again, conference officials and the government representatives there refused to adopt the notion of water as a human right, or to address the growing criticisms of the private water experiment.

The World Bank chose to use Kyoto to launch a major report on water financing it had been working on for some time. Reaction to the private water companies was growing in the global South; the big water companies were worried that their presence was not sustainable without guarantees from the international financial institutions that they would be protected from both local political turmoil and currency crises in the countries of Latin America in particular. *Financing Water for All* was written by a panel chaired by Michel Camdessus, formerly head of the International Monetary Fund and then honorary governor of the Banque de France, and included Gerard Payen, former executive vice-president of Suez; Charles-Louis de Maud'huy, a Veolia director; and Ismail Serageldin, then with the World Bank.

The Camdessus report was everything the water companies wanted to hear and provoked a strong negative reaction both at the forum and internationally. It recognized that corporations operating in the developing world were experiencing strong resistance and said they had to be protected, both politically and financially, with major new investments of US$180 billion. The panel called for full-cost recovery for water projects and advocated the use of public funds to pay for the preparation of private contracts and tenders. It also called for a Liquidity Backstopping Facility to guarantee corporate profits in cases of currency devaluation and political conflict. The clear message was that, without more public financing, the big companies could not guarantee a continued presence in poor countries. Governments bought the recommendations and took them home to become part of development programs.

Fourth World Water Forum – Mexico City, March 2006
The Fourth World Water Council in Mexico City in 2006 was also very large, with 20,000 delegates, 1,500 journalists and government representatives from 140 countries. But by this forum, the water policies of the World Bank had become so unpopular, a legion of armed guards and police had to protect delegates behind a massive security wall. That this US$220 million extravaganza was really a corporate trade show was publicly revealed, and organizers were castigated for charging US$600 per person to attend. Its corporate sponsors included Coca-Cola and Grupo Modelo, the Mexican beer giant.

Both because of the blatant corporate nature of this gathering as well as the fact that the World Water Council clearly does not want a meaningful dialogue with civil society, many NGOs boycotted the Fourth World Water Forum, choosing instead to convene their own assembly of one thousand delegates. They also organized a massive rally of forty thousand protestors, who took over the streets of downtown Mexico City, chanting "Our water is not for sale" and calling on governments to leave the World Water Forum and join their citizens on the street.

Water Privatization Has Been a Complete Failure

Almost twenty years of documented cases of the failure of privatization and growing opposition to the World Bank and the water service companies in every corner of the globe have revealed a legacy of corruption, sky-high water rates, cutoffs of water to millions, reduced water quality, nepotism, pollution, worker layoffs and broken promises. The reality is that for-profit companies, even if operating honestly, cannot practice desperately needed water conservation and source protection. In fact, to stay competitive, water companies are relying on deteriorating water quality around the world.

Nor can competitive corporations supply water to the poor. This is and will remain the role of governments. The ultimate goal of private companies is to make a profit, not to fulfill socially responsible objectives such as universal access to water. In countries where most of the population earns less than two dollars a day, notes Sara Grusky of Food and Water Watch, private companies cannot meet shareholder obligations to provide a market rate of return. Nor can they expand their services to a population that cannot pay. The only way that the private sector can stay competitive in such a situation is to have access to public subsidies, the very thing they were supposedly brought in to relieve. In fact, in most cases, the promise that the private sector would bring needed efficiencies, expertise and new investments never materialized.

In *Pipe Dreams: The Failure of the Private Sector to Invest in Water Services in Developing Countries*, a groundbreaking report published in 2006 by Public Services International and the World Development Movement, authors David Hall and Emanuele Lobina clearly demonstrate that the investment argument of the World Bank is a myth. In all of sub-Saharan Africa, South Asia and East Asia (excluding China), only about six hundred thousand new connections to households were made as a result of investment by private sector operators since 1990, extending services to only around three million people, a mere fraction of the numbers targeted by the UN. Even this small number must be offset, say the authors, by the number of households who have been cut off due to non-payment of water bills and the fact that most of these hookups were also state-subsidized. As well, even most of these "success" stories failed to deliver the investment and extensions promised when the contracts were originally set up. The only part of the world where the private sector has contributed more to the extension of water connections is Latin America, and research shows that these achievements were no better than those of the public sector in general, and worse in several important cases.

Perhaps even more damning is the fact that, anticipating the private sector would be bringing in new infusions of money, the World Bank, regional banks and First World country donors actually *decreased* their funding of water services in the global South. Between 1998 and 2002, investments in water infrastructure in developing countries by donors and development banks fell from US$15 billion to US$8 billion. At the same time, World Bank policies themselves discouraged poor countries from investing in services such as water. "The net contribution of 15 years of privatization has thus been to significantly reduce the funds available to poor countries for investment in water," says the report. "The focus on private sector development has contributed to a reduction in the level of aid and development finance from donors which is far greater than the actual investment made by the private sector." Further, the World Bank punished countries that resisted privatization with reduced support. Perhaps most disturbingly, the authors contend, is that the big water corporations have become so powerful, they actually influence which countries, regions and cities will receive investment funds from the global North. Because the decisions are based on where the corporations can make a profit, the most desperate communities have not received their share of funding.

The January 2007 United Nations Development Program report of the International Poverty Centre confirms these findings. Kate Bayliss of the United Kingdom and Terry McKinley from Brasilia find that, because it is so poor, sub-Saharan Africa received only 4 percent of global private investment between 1990 and 2003. In order to attract more investment, poor countries have had to realign their expectations and focus on creating a favorable climate for business instead of worrying about delivering programs to the poor. The study also confirms that initial hopes for privatization were so high, donor spending on infrastructure fell in the expectation that the private sector would take up the ensuing slack. In 2002, World Bank lending for water and sanitation in sub-Saharan Africa was only one-quarter of what it had been in the period 1993–97. At the same time, the World Bank increased its support for private investment through its International Finance Corporation and its Multilateral Investment Guarantee Agency. "Hence," say the authors, "African countries have been caught in a terrible bind. Not only has donor financing of public investment declined but also private investment has followed suit."

Many new studies and reports repeat these findings. According to an April 2006 report, published by the Norwegian Forum for Environment and Development, water privatization has failed to deliver water to the poor; undermined the human right to water; taken place at the expense of democratic principles and with minimal accountability to local citizens; and has led to foreign control of water and the creation of monopolies. A September 2006 report by the German-based Institute for Development and Peace critically examined the impact of water privatization on development and called for state regulation and compulsory ethical guidelines for all water investment projects.

"There has been bribery, corruption, non-compliance with contractual agreements, lay-offs, tariff increases, and environmental pollutions," concludes Naren Prasad, research coordinator for the United Nations Research Institute for Social Development. "'Sign and renegotiate' is the order of the day, and the World Bank has even published a manual on how to renegotiate a failed concession contract." Prasad notes that the World Bank is trying to repackage and relaunch privatization under the "softer" guise of public-private partnerships.

Even those traditionally supporting water privatization have now distanced themselves. In September 2006, WaterAid, the British water charity started by the private British water companies, issued a strong condemnation of the European Union's promises to supply water to the world's poor. "Not an additional person" has benefited from the European Union's Water Initiative (EUWI), said WaterAid, noting that the share of European aid for water projects had fallen from 5.5 percent in 2000 to 4.2 percent in 2003. "Despite the proven disinterest of international investors in financing water and sanitation projects in developing countries, the EUWI persists in trying to attract private money, leaving no opportunity to debate the need for increased EU aid to the water sector."

At the Fourth World Water Forum, the United Nations released the *UN World Water Development Report 2006*, with a highly critical analysis of water privatization. Noting that many water companies were retreating from the global South because they were encountering resistance, the report castigated them for pulling out of non-lucrative contracts, saying that millions of people were going to have to wait for services as a result of their retreat. The UN also noted that those who have benefited from private water services in the Third World come almost exclusively from the wealthy sectors of society. "To provide the poorest section of society with adequate water services is typically viewed as a high risk enterprise that largely lacks opportunities for economic return," said the UN, adding, "It is high time to bring the governments back in."

Undeterred, the World Bank published a "Toolkit on Water and Sanitation" in 2006 to "assist governments in developing countries that are interested in using private participation to help expand access to water and sanitation." The Toolkit will "review key issues that governments need to resolve in order to introduce private participation; . . . consider some of the key reform choices for the water sector upstream of the private participation arrangements"; . . . and "consider how the chosen arrangement can be embodied in legally effective laws, contracts and licenses." Michael Goldman explains this kind of intransigence in the face of contrary evidence as one of ideology: "The World Bank's policy campaign for water privatization has been much more than a leasing program for dilapidated public plumbing and sewer infrastructure. Rather, it has marked the entrance of new transnational codes of conduct and procedures of arbitration, accounting, banking, and billing; a whole new ethics of compensation (including the ability of foreign corporations to sue a government); new expectations of the role of the public sphere; and the normalization of transnational corporations as the local provider for public services and goods."

The Big Water Service Companies Cash In

Notwithstanding the mistakes they have made and the resistance they have encountered, the water service companies, especially Suez and Veolia, have continued to make large profits. In 1990, only a fraction of the world's population, about fifty million people, bought water from a private water provider. Today, big water companies provide water to about six hundred million people, 10 percent of the world's population, a huge increase in a short time. (This does not mean that 90 percent of the world is serviced by the public sector. There are still almost one and a half billion people with no water service at all – public or private. It is more accurate to say that, of the four billion people who have water services, around 15 percent are now buying their water from the private big water operators.) The water companies themselves conservatively estimate that within ten years, the number of people buying their water will double. This has provided a bonanza for the water industry.

Suez, number seventy-nine on the Fortune 500 list of corporations, has 160,000 employees worldwide – 72,000 of whom now work just in its water division – and revenues of almost

US$60 billion. Veolia Environment has 272,000 employees, 70,000 of whom work just in water, and revenues of just under US$34 billion – up from just US$5 billion a decade ago. Suez posted a healthy 6.7 percent revenue increase in 2006, although its water division's revenue growth was more modest, at 3.2 percent, and Veolia's revenues rose almost 12 percent, a profit increase of 55 percent. Between them these two transnationals have, until very recently, controlled two-thirds of the global private water services sector. In third place is Thames Water (recently divested by RWE) with 12,000 employees and revenues of more than US$2 billion. SAUR, of France, and Agbar of Spain, are the fourth and fifth largest water companies in the world. Other big players include AquaMundo of Germany and Biwater, Severn Trent, Kelda Group and Anglian Water of Great Britain.

It is true that the companies have met with some very hard resistance in communities around the world (see Chapter 4) and consequently have pulled out of many contracts. Suez, for instance, announced in its 2006 annual report that its withdrawal from the non-lucrative contracts of Latin America, sub-Saharan Africa and South Asia is pretty well complete. As well, several big players are backing out of the water business altogether. RWE, the German energy giant, divested Thames Water after several years of intense local resistance, particularly in the United States. Its rate of return on investment was well below the company standard, and it was just not profitable for the company to stay in the water business. Enron divested itself of its water company, Azurix, in early 2001 (just months before Enron's scandal broke and it was forced to declare bankruptcy), when it became apparent its foray into the water business had been a complete failure. Giving new meaning to the word *hubris*, Azurix's president, Rebecca Mark, had just declared that she would not rest until all the world's water had been privatized.

Nevertheless, it is premature to cite the demise of private water services or to underestimate the staying power of these global competitors. New World Bank data show that the number of new public-private contracts awarded in 2005 was the highest since 1990, and that 2006 would likely be another banner year. As well, reports *Global Water Intelligence*, merger-and-acquisition activity in the water sector has been remarkably intense in the last decade, with the private sector acquiring close to US$10 billion in equity stakes from governments and municipalities. As well, private utilities are being subsidized by the public pension funds of the rich global North. In 2005, U.K. water utility AWG was acquired by a group of pension funds from Canada and Australia in a deal worth more than US$4 billion. Thames Water was sold to Kemble Water Limited, a consortium led by Macquarie's European Infrastructure Fund for close to US$5 billion.

What is changing, reports *Masons Water Yearbook*, the industry bible, is that worldwide, a whole new raft of water companies is entering the market, actively competing with the big two. In fact, the total market share of the big five combined fell to 47 percent in 2006, an astonishing change in just two years.

Further, these big competitors, as well as dozens of smaller ones, are no longer concentrating solely on the global South, attempting to get into lucrative markets in more "stable" countries, with some real success. In 1999, Vivendi Environment – now Veolia – bought U.S. Filter for US$ 6.2 billion and became a member of the powerful U.S. Coalition of Service Industries. The next year, Suez bought United Water for US$1.2 billion cash and in February 2007 added Aquarion–New York, one of the most important water distribution systems in the United States, to its acquisitions. And in 2003, RWE Thames bought American Water, which served 15 million people in twenty-seven states, for US$8.6 billion. With these purchases, the three biggest American private water companies were taken over by European water conglomerates whose goal is to control 70 percent of the U.S. market within the next two decades. Soon, the big three were running the water systems in Atlanta, New Orleans, Tampa, Indianapolis, Oklahoma City, Stockton, Milwaukee, Springfield, Pittsburgh and Honolulu, to name just a few.

They brought with them a new level of political lobbying, which they had learned in Europe. Before the big European water companies took over the industry, reports the Center for Public Integrity, the water utilities spent very little on political contributions. From 1995 through 1998, they spent less than half a million dollars on campaign contributions. But in the elections of 2000 and 2002, campaign spending more than tripled to roughly US$1.5 million. More than half of this money came from just two companies, United Water and American Water, both of which are owned by foreign water corporations. Taking advantage of changes to federal laws in the 1990s that require utilities to consider private partnerships with water companies before they receive federal assistance, the private companies had already doubled the rate of privatization of water services during the 1990s.

Now, armed with new liberalized federal tax laws that allow municipalities to enter into long-term private water utility contracts of up to twenty years, they were poised to make another breakthrough into billion-dollar contracts. The number of water systems operated under long-term contracts by private companies grew from approximately four hundred in 1997 to about eleven hundred in 2003. This type of contract makes it far more difficult for a city to cancel if it decides that privatization has been a mistake. Today, the National Association of Water Companies reports that in 2006, the private water utilities market in the United States, with more than thirteen hundred public-private partnerships, is a US$1.7 billion business and supplies 15 percent of Americans with water services.

The water companies are now pushing for legislation in the United States to require cash-poor municipal governments to consider privatizing their water systems in exchange for federal funding for water delivery. And they are keenly supporting a Bush initiative called the Water Enterprise Bond that would remove water and wastewater utilities from state volume caps, making unlimited private-activity bonds available. The initiative is supported by the Environmental Protection Agency. "We believe this is a significant step . . . that can bring potentially billions of new dollars for water and wastewater infrastructure," Benjamin

Grumbles, assistant administrator for the EPA water office, told an EPA-sponsored conference on water infrastructure in Atlanta, Georgia, as reported in the March 22, 2007, *Bond Buyer*.

Europe is another region targeted for growth by the big water companies who are working with allies on the European Commission to foster a positive climate for public-private partnerships in countries that have adamantly refused to consider private water systems. About 70 percent of Europe's water services are currently in public hands, but there is growing pressure to allow the big water companies to compete for contracts. The "big two" are particularly targeting Germany, Austria and Italy as politically ripe for privatization. France, Spain, Wales and England are already largely privatized. In several major reports, the European Commission has recently recommended public-private partnerships as the way forward for a continent in need of major infrastructure investment.

David Hall of the Public Services International Research Unit warns that the companies have also targeted the poorer countries of Eastern and Central Europe where the big water corporations, with the support of the European Bank for Reconstruction and Development and the World Bank, are active in Croatia, Albania, the Czech Republic, Romania, Serbia, Estonia and Hungary. *Masons Water Yearbook* reports that in 2005, the water business in Europe "surged." The Middle East is also a target market. Saudi Arabia, for example, began privatizing water services only in 2006 but expects that by 2010 private companies will provide water for half the population.

But the biggest plum of them all is China. Suez Environment, Veolia and Thames are all very active in China, where they began to spread their empires twenty years ago. They used to "produce" and sell water in bulk to municipalities because the Chinese government kept strict control of water delivery to their populations. With the new market-friendly regime in China, however, the companies can now obtain long-term distribution contracts. Veolia provides full water services in seventeen cities, including Shenzhen, Kunming, Shanghai and Changzhou; Suez does the same in Chongqing, Sanya, Tanggu and Tanzhou. Debt collection, with a 99 percent payment rate, is heavily enforced by the government.

The profits are astounding by any measure, generating returns of more than 20 percent on average. In 2002, Thames Water acquired the largest single shareholding in the China Water Company; the purchase boosted Thames's customer base in China to 6.5 million. Suez currently has nineteen joint operations in sixteen cities – an investment of US$640 million – and will double its investment in the next two years. Suez Environment saw a 26 percent growth in mainland China in 2006 and recently moved its Asia-Pacific regional headquarters to Shanghai.

~~~

The big water utilities are keenly aware that the water business landscape is shifting under their feet and are poised to compete with a whole new set of competitors entering the newest – and hottest – market in the world, the water reuse industry. However, their lasting legacy is likely that, in league with the World Bank and the United Nations, Suez, Veolia and the other big water service companies set the stage for the complete commodification of the world's water and the conditions for the creation of a global corporate-owned water cartel.

Chapter 3

The Water Hunters Move In

These are ugly decisions, but you either drink water or you die.

– Peter Beattie, premier of Queensland, Australia, in response
to public opposition to drinking recycled sewage water

The big water utility companies, such as Suez and Veolia, while still powerful and very wealthy, are facing heavy competition from a host of new companies entering the fastest-growing arenas of the water market, namely, providing clean water for industrial and municipal use and cleaning dirty water when that use is done. The industry's technology section is growing at twice the rate of its utility section and already accounts for more than one-quarter of all revenues. Global water reuse capacity will rise by 181 percent over the next decade, according to a *Global Water Intelligence* report, *Water Reuse Markets 2005–2015: A Global Assessment & Forecast*. Water reuse investment will total US$28 billion. Industry heavyweight Siemens estimates that the recycling technology market alone is currently worth US$40 billion and will double in the next eight to ten years.

William J. Roe, chief operating officer for Nalco, a global water-treatment giant whose water division employs 10,000 people and operates in 130 countries, said that his company believes future wars will be fought not over oil but water. According to Andrew D. Seidel, chief executive of U.S. Filter, a Palm Desert, California, utility that is getting into the water treatment business, the water business has at least 100,000 players providing "all kinds of products and services."

"Water companies provide the chemicals that purify water used to make computer chips – water too pure for ordinary human needs because it would leach calcium, zinc, and other vital minerals out of the body," reported Claudia H. Deutsch in the *New York Times*. "They sell the additives that keep factory pipes safe from corrosion when water cools them down or heats them up, and that make saltwater as sweet as liquid from freshwater streams. They ensure that Coke or Starbucks coffee tastes pretty much the same, no matter where you order it. They enable refineries to turn cheap, highly acidic crude oil into high-grade gasoline. Lately, they are even helping hotels, hospitals, and apartment buildings keep micro-organisms like the ones that cause Legionnaires' disease from circulating through plumbing and climate-control systems."

Masons Water Yearbook confirms this trend. Desalination and industrial water outsourcing are the global water industry's most dynamic sector, it reports, and will grow at a healthy rate for many years. This trend will mean that the big European water companies will no longer dominate the water market, notes *Masons*, which documents the emergence of major new water corporations in Morocco, Poland, Russia, Sweden and the United States. "Water is hot," said Debra G. Coy of Schwab Capital Markets at a January 2004 conference held by the National Council for Science and the Environment. Vast spending required to meet demand for new supplies and to upgrade deteriorating or nonexistent water infrastructure around the world is creating a global water industry, she added, and technological solutions are generating untold investment opportunities.

Saudi Arabia plans to spend us$80 billion to increase water production capacity in desalination and other water purification technologies over the next two decades. In the next decade, Dubai will invest us$100 billion in water treatment technology. China, now fueling 30 percent of the world's economic growth, is planning to add between 200 and 400 power plants with an approved 300 gigawatts. However, it does not have the water available for their construction or operation. As well, the Ministry of Water Resources forecasts that China's population will soar to 1.6 billion by 2030, requiring an increase of 130 to 230 billion cubic meters in water supply availability. So the Chinese government has earmarked US$125 billion toward improving water quality between 2006 and 2010. Suez's wastewater engineering treatment subsidiary, Degremont, has already built 160 water plants in China and plans to build another 160.

The global demand for infrastructure construction and repair could reach US$20 trillion in the next twenty-five years, says John Balbach, partner at Cleantech Group, a venture capital research firm based in Michigan. According to the Congressional Budget Office, the United States alone will have to invest close to US$40 billion a year for at least a decade to rebuild old pipes and other water infrastructure on top of the US$55 billion annually it will cost just to maintain and operate the system. The American Society of Civil Engineers says that upgrading the municipal water system in the United States could cost US$1 trillion over the next several decades. This story is repeated around the world as governments are under pressure to upgrade antiquated systems leaking massive amounts of water through old pipes. As well, the market for private industry contracts is exploding. The industrial water outsourcing market offers great opportunities because industrial clients prefer to pay water and wastewater services companies to help them comply with environmental legislation rather than establishing in-house treatment capabilities.

New Water Companies Enter the Market

U.S.–based ITT Corp. claims to be the world's largest provider of water and wastewater treatment systems (so do Nalco and U.S. Filter) with operations in 130 countries. ITT's water division is diverse and creative: for instance, it runs the pumps that drain the Hoover Dam's interior tunnels as well as 250 wastewater treatment stations in China. Its product lineup includes pumps for residential, municipal and commercial water systems; biological filtrations and disinfection treatment for municipal and industrial wastewater; and pumps for mining, chemical, paper and petroleum factories.

Another American company – Danaher Corporation – a US$9 billion manufacturing and tool company, has also entered the water treatment business in a major way. These two join a whole host of new U.S.-based water companies eager to cash in on the private water industry, now estimated by *Environmental Business International*, the online research source for the sector, as a US$100 billion industry in the United States alone.

Global giant GE has also recently entered the market, raising the competition stakes. In 2001, the company bought Betz Dearborn, a major U.S. player in water treatment chemistry, and two years later, it purchased Osmonics, which makes membranes for use in water treatment, thus launching GE Water Technologies, now worth US$1.4 billion. That same year it bought Ionics, a U.S.-based desalination plant for US$1.1 billion, as well as Zenon Environmental Inc., which make ultrafiltration membranes. Company water researchers use imaging technology borrowed from GE Medical Systems to "diagnose" chemical problems in water used in manufactured goods as diverse as cars and pipes. GE has stated it wants a US$50 billion slice of the global water market. In February 2007, GE announced plans to invest more than US$1 billion in European infrastructure development with an eye to becoming a serious competitor in that market.

GE, ITT and Nalco are now the big guns in the field, with more than 40 percent of the market, but other key players are entering the field as well. Dow Chemical recently launched Dow Water Solutions to create "safer, more sustainable water supplies for communities around the world." Its water treatment and desalination division has become the fastest-growing part of its business and earned US$450 million in 2006 alone. (To counter criticism that the company that brought the world napalm and Agent Orange should stay out of the water business, Dow is financing the annual Blue Planet Run, which sends runners around the world to raise money for safe water projects in the Third World.)

Of course, the utility giants are not to be written off. The two largest waste management companies in Europe are SITA, the waste services division of Suez, and Veolia Environmental Services (formerly Onyx), the waste services division of Veolia. The third is Remondis, formed when its parent, Rethmann, took over RWE's waste division when the company spun off its water sector. They are all chasing water treatment contracts. In 2004, Veolia won a US$1.3 billion contract from Peugeot Citroën in Eastern Europe to clean up the water used in its car production and in 2006 signed a contract to operate facilities to collect, treat and recycle industrial wastewater at the Beijing Yanshan Petrochemical site. Suez shares leading-edge biological and chemical water treatment technologies in China from its R&D center in Paris. Steve Clark, executive director for Suez Environment in China, told *EuroBiz* Magazine that the company would soon open a multimillion-dollar water R&D center in Shanghai.

The water industry is beginning to create "centers of excellence" in certain universities and regions. Singapore is becoming a "global hydrohub," exporting private water sector experts and technology around the world. Singapore's Environment and Water Industry Development Council announced in February 2007 that it is setting aside US$330 million to fund world-class water research centers and train researchers in water solutions for the private sector. Black & Veatch, a major engineering and consulting company with ninety offices around the world and one of the "Top 500" private American companies, has moved its global water business to Singapore to take advantage of this opportunity. Black & Veatch has designated Singapore as its Global Design Center as well as its Center of Excellence in Desalination and will "grow" the number of technical water experts working for the company tenfold by 2009.

German industrial giant Siemens is another major player in the global water business. In 2004, the company bought U.S. Filter from Veolia for US$1 billion (U.S. Filter claims to be the "world's largest and fastest-growing water and waste water treatment company") and has partnered with the Israeli water utility Mekerot. With close to six thousand employees in its water division, Siemens Water Technologies, the company announced in late 2006 that it is setting up a US$50 million global water and research development and engineering center in Singapore as well. And Singapore is creating its own competitors. A host of new Singapore water treatment companies, including Sembcorp, Dayen, Darco, EcoWater, Salcon and Hyflux, are fanning out across Asia and China. (Hyflux is the company that developed controversial NEWater, totally recycled sewage water that the Singapore government uses to supply its population with drinking water. The company is now working with the Public Utilities Board in Australia to sell NEWater to skeptical Australians and is poised to grow in India, Thailand, the Middle East and China.)

From Nuclear to Nano, Nothing Is off the Table

Desalination

In spite of the documented problems with desalination, all indications are that this sector of the water market is about to take off. According to *Global Water Intelligence*, the global desalination industry is set to almost triple by 2015. This expansion will entail capital investments of close to US$60 billion in those years. (*Water Industry News*, the online journal of Environmental Market Analysis, puts this figure higher, at US$95 billion.) Of that, more than half is expected to come from the private sector, meaning that desalination is more open to private sector participation than any other part of the water business. Desalination is also the most high-tech and most international part of the water industry. The market is expected to reach US$66 billion in 2010 and US$126 billion by 2015.

The largest market for desalination will continue to be the Gulf area, which will double its capacity in the coming decade. The largest growth area will be the Mediterranean Rim, where Algeria, Libya and Israel are anticipating capacity increases in excess of 300 percent. The United States will "make the breakthrough," says *Water Industry News*, into large-scale municipal desalination, as will China and India. Sydney, Australia's planned plant will likely be run by the private sector on a for-profit basis, the government has confirmed.

There are around eighty-seven desalination corporations in the world, and that number is steadily rising. Utility leaders Suez and Veolia have both entered the market as have water purification and wastewater leaders ITT, Siemens, Dow, Nalco and GE, which operates the world's largest "mobile desalination fleet" for emergencies and supplies water for hotels and resorts around the world. GE has joined forces with another American company, Pall Corporation, the largest producer of membrane filtration systems in the world with close to US$2 billion in sales in 2006, to expand into the global desalination market. GE has also teamed up with Saudi Arabia's chief desalination corporation, ACWA, as well as Suez, to build new desalination plants in that country. In February 2007, GE announced that it will build a major new desalination plant, Africa's largest, in Algeria. Thames Water is seeking permission to build a desalination plant to convert salty water from the tidal Thames River in East London. But London mayor Ken Livingstone is opposed. For years, he has been battling with the company, which provides private water services to the city, to fix the old pipes that leak close to a billion liters of clean, purified water every single day. The mayor points out the irony of a private company wanting to make a profit from desalinating brackish water supposedly needed because of a water shortage – one the company could rectify if it stopped wasting volumes of clean water every day.

Other major players include Spain's Befesa, which in early 2007 completed financing to design, build, own and operate for twenty-five years a new desalination plant in Chennai City, Tamil Nadu, the largest desalination plant in India. Another major Spanish player is Inima, which owns and operates twenty-five desalination plants around the world. Consolidated Water is registered in the Cayman Islands and operates plants in Belize, Barbados, the British Virgin Islands and the Bahamas. Metito operates in fourteen Arab and Middle Eastern countries and has expanded to countries as far away as Argentina and Australia. Israel's IDE is becoming a global player, having built the largest plant in Israel and now expanding into the Mediterranean.

The major U.S. companies are Stamford, Connecticut–based Poseidon Resources, which built the Tampa Bay desalination plant, the largest in the western hemisphere, and CalAm (California American), the largest privately owned water company in the United States with subsidiaries in twenty-seven states. RWE now wholly owns CalAm, in the wake of the 2003 takeover of the company by its subsidiary, American Water (which it is trying to divest). CalAm is currently vying to open a controversial desalination plant near the Monterey Bay National Marine Sanctuary.

Then there are young upstarts, such as Aqua Genesis, a Las Vegas company that has invented the Delta-T – a desalination device that can be powered by geothermal heat so it can work even if the power grid goes down. Ronald Newcomb, company co-founder and director of operations at the Center for Advanced Water Technologies at San Diego State University, says of his company's prospects, "We tried to estimate the potential size of our company, but the numbers became so big, we just stopped."

Nuclear Desalination

Increasingly, because of the high cost of conventional desalination technology, the notion of using nuclear reactors to provide the needed energy is taking root in many important circles, including the International Atomic Energy Agency, which won the 2005 Nobel Peace Prize. Its Desalination Economic Evaluation Program (DEEP) defines nuclear desalination as the production of potable water from seawater in a facility in which a nuclear reactor is used as the source of energy and the agency openly promotes the technology internationally. The industry has a new journal, the *International Journal of Nuclear Desalination*, founded in 2004 with an impressive international board of academics and representatives of the atomic energy agencies of many countries. The American Nuclear Society supports nuclear desalination, as does the European Nuclear Society.

Nuclear desalination plants are already operating in Japan, Kazakhstan and India. New nuclear desalination plants are slated for India, Pakistan, Egypt and China; and Russia, Morocco, Tunisia, Australia, Algeria, Iran, Indonesia and Argentina are all seriously considering the option. *Global Water Intelligence* reports that nuclear desalination will play an increasingly important role in these countries and is increasingly supported by their governments. Texas senator Kay Bailey Hutchinson supports a University of Texas initiative to conduct scientific research and train nuclear engineers toward the creation of desalination technology. The project is to be a state-of-the-art, helium-cooled nuclear research facility built largely underground in Andrews County near a proposed uranium enrichment facility.

As momentum for nuclear desalination grows, so, too, do market opportunities for the private sector, not only for the players currently in the water business but a whole host of new players, from uranium companies to the nuclear industry. In March 2007, Suez CEO Gérard Mestrallet announced his company's intention to invest in new nuclear capacity in the "timeframe 2015–2020." Although Suez is targeting nuclear-powered electricity, it would not be a stretch for the company to design and invest in nuclear powered desalination plants for its water division.

Nanotechnology

Nanotechnology is a new field of applied science and technology dealing with engineering at the molecular scale. Nanoparticles are named for their small size – a nanometer is a billionth of a meter – and are smaller than anything humans have ever put into commercial products before. Nanotechnology has been touted as the next industrial revolution and is now being used in products as diverse as sunscreen and stain-resistant clothing to food, food packaging, dietary supplements, electronic equipment and drugs. Most of the big food and drug companies have been investing heavily in nanotechnology research for at least a decade. According to respected physical science journal *Lux Research*, the industry will grow to US$2.6 trillion in manufactured goods alone in the next five years. Drug companies are earning almost US$1 billion a year already on the technology. The U.S. Patent and Trademark Office has issued more than four thousand nanotechnology related patents already and twenty-seven hundred are pending.

More recently, the big global desalination and water purification and treatment companies have set up divisions to explore the latest technological superstar in the water business: water nanotechnology. In order to clean dirty water, company scientists are looking to the sub-microscopic world of nanoparticles to seek and destroy the sources of groundwater pollution with various forms of water nanotechnology such as nanomembranes and nanoporous zeolites (microporous crystalline solids). They are aided in their quest from extensive research largely funded by governments. For instance, the Israeli government is heavily investing in nanotechnology research at five universities – the Weizmann Institute of Science, the Technion, the Israel Institute of Technology, Tel Aviv University and Bar-Ilan – and the U.S. government is investing about US$2 billion on research and development of nanotechnology.

Various U.S. government departments came together in 2004 to form the National Nanotechnology Initiative, which more recently began to fund university research in water nanotechnology. Much excitement was generated with the November 2006 announcement that scientists at Rice University in Texas, using nanoparticles of iron oxide five thousand times smaller than the width of a human hair, had found that precision-engineered particles were capable of stripping far more toxic contaminants such as arsenic from water than existing filters. That same month, researchers at UCLA announced the development of a new reverse-osmosis membrane that they claim will reduce the cost of seawater desalination and wastewater reclamation. In early 2007, the Environmental Protection Agency announced ten research grants worth US$5 million to find nanotechnology solutions for poor-quality drinking water.

However, as in so many other cases of government-funded university research, it is the private water companies that are taking control of this technology as well as advantage of the funded research, sensing the enormous potential for profit. The UCLA scientists are working with a private California nanotech firm called NanoH2O to develop a patent on their discovery. They expect their new membranes to be commercially available sometime in 2008. The company says right on its website that its products are based on the UCLA research. The UCLA scientists co-founded the Water Technology Research Center in 2005, which collaborates with UCLA's NanoSystems Institute, whose mission is "to encourage university collaboration with industry and to enable the rapid commercialization of discoveries in nanosystems."

Titanium dioxide technology is being tested and presented to industry right now. The Department of Energy's Pacific Northwest National Laboratory, whose mission is to "move science to the marketplace," says on its website, "These novel materials are ready to be commercialized." Dozens of other small water nanotech companies are springing up in the United States, including KX Industries in Orange, Connecticut, Argonide, in Sanford, Florida, and eMembrane in Providence, Rhode Island. The story is the same in many other countries. Australia hosted a June 2006 forum called Commercialising Nanotechnology in Water, which brought together leaders in government, academia, science and industry to "aid the development and commercialisation of water nanotechnologies." The major outcome of the conference was the creation of a National Consultative Committee on Nanotechnology Commercialisation to "coordinate and accelerate innovation and serve as a platform for Government, Industry and Academia to drive the adoption of commercial nanotechnology."

Of course, Suez, Veolia and GE are already the leaders in water-based nanotechnology. Veolia has teamed up with Dow Chemical subsidiary Filmtec to develop its own nanofiltration technology. Suez has installed nanotechnology "ultrafiltration" systems in a plant outside Paris. GE's Water Technologies mission is "to be recognized as the world's best supplier of engineered and mechanical treatment programs for water" using nanotechnology. The German chemical groups BASF's future business fund has devoted a significant proportion of its US$105 million nanotech research fund to water purification technology. Ever on the leading edge of coming technology, these and other private corporations are moving in to take control of the hottest new technology around.

Emerging Technologies

Then there are the emerging technologies creating investment interest. These include the manufacturers of Atmospheric Water Generators (AWGS) – machines that literally suck water from the air. There are dozens of manufacturers of small machines now supplying homes and offices such as China-based Hendrx Corp., with annual sales of more than US$5 million, and Singapore's Hyflux. But some are aiming to take larger amounts of water from the atmosphere to supply dry regions and populations with water. Aqua Sciences is a Florida-based AWG company that has developed the technology to produce forty-five hundred liters of water a day from fully self-contained mobile freshwater generating systems (which look like large trailers) now under contract to the Pentagon to provide water to American soldiers fighting in Iraq. Free Water Inc. is another U.S.–based AWG company producing large volumes of water from the atmosphere. Ironically, given its name, Free Water developed patents with its partner, Air2Water, that will allow it to "control and develop the market without competition."

Cloud seeding – the practice of seeding clouds with silver iodide and dry ice from airplanes in order to enhance the possibility of rain – is growing and is currently used in twenty-four countries, including Australia, the United States and China, where the practice is so common, there are now conflicts between villages and cities over "cloud theft." China is the world's leading cloud seeder, spending about US$50 million a year and employing thirty-five thousand people to take the rain from clouds. Officials estimate that cloud seeding has increased rainfall by 10 percent, but some scientists worry that the practice may be having detrimental effects on the hydrologic cycle. While the companies in the cloud-seeding business are still mostly local, this market clearly has the potential to grow.

Water Property Rights

There is also the new practice of buying, trading and selling bulk water and water rights. A slew of private water brokerage companies with names such as Watermove, Waterfind and Elders Water Trading sprang up in 2001 after the government of Australia changed the law to allow rural landowners and farmers to "delink" water from the land and sell it to urban dwellers. WaterBank is a New Mexico–based brokerage and investment-banking house that "connects sellers of water rights, water utilities, springs, geothermal water and bulk water with buyers." The company claims to have 375 sources of water for sale worldwide but gives very little information on its website about the actual company and its directors. In explanation, WaterBank says, "WaterBank and its staff carry out a significant amount of investigative reporting that is uniquely reported on this web site. Because of the highly political character of water, and potentially dangerous agendas of the actors, we consider ourselves as journalists, and as such, the sources of much of our material is [sic] strictly confidential and must remain confidential."

WaterColorado.com, a Colorado-based water trader, reports that water prices jumped 40 percent in the state in 2006 alone. Company president Joe O'Brien says that in the 1950s, an acre-foot of water from the Colorado Big Thompson project was worth a dollar and is now selling for US$16,000. (One acre-foot of water is equal to about 1.3 million cubic meters and can support two families of four for one year at current U.S. use.)

Perhaps in anticipation of this emerging water market, water entrepreneurs are hunting for new sources of water and buying up bulk water and water rights and holding them for future profit. Vidler Water Company is the water brokerage arm of PICO Holdings, a California water resource firm in the business of "the acquisition of strategic water resources and water storage facilities" in order to "establish a long-term income stream through the sale or lease of water resources and underground storage facilities to both public and private users." Dorothy Timian-Palmer, former water manager for Carson City, now Vidler's president, calls her company a "water developer." As of early 2007, the company owned more than 135,000 acre-feet of water rights in Nevada and Arizona – currently worth about US$500 million. But the company is holding on to most of its water and planning to buy up more because the price of water is steadily and inexorably going up in the American Midwest.

T. Boone Pickens, the seventy-eight-year-old Texas oil billionaire, has started a company called Mesa Water, which has bought up 200,000 acres (about 80,000 hectares) of groundwater rights in Roberts County, from which he expects to make more than US$1 billion on a US$75 million investment. He also is seeking the right to pump Ogallala water and sell it to El Paso, Lubbock, San Antonio and Dallas. So far, his water hoarding has gone unchallenged by state authorities.

Not surprisingly, several companies are ready to ship bulk water around the globe for a price. World Water SA is registered in Luxembourg but operates out of Anchorage Alaska, where its president, Ric Davidge, lives. Davidge, the former water czar for the Alaskan government, paved the way for bulk water sales. He now heads World Water, which identifies "viable water markets around the globe to enter into long-term take-or-pay bulk water contracts with public or private buyers." His company partners with the Japanese shipping company NYK; Nordic Water Supply, which runs a water bag transfer operation between Cyprus and Turkey; and Alaska Water Exports, which plans to export bulk water from Alaska to California.

South Carolina–based Flow Inc. wants to ship water in empty ballast tanks on big oil tankers after they have dumped their cargo at U.S. ports and return to the Middle East. Company president Eugene Corrigan says that he is on the verge of signing a contract with a major tanker company that could eventually export 165 million cubic meters of freshwater a year to thirsty Middle Eastern countries that can pay for it. He told *Global Water Intelligence* that the wealthy Gulf States already have the infrastructure to offload bulk water shipments.

The Bottled Water Industry Is Thriving

Then, of course, there is the phenomenon of bottled water. Bottled water isn't new, although it was originally created as a curative for the wealthy. In 1855, France's Vittel Grande Source got a permit to sell its mineral water in individual containers; a few years later Perrier received a similar license. A hundred years later, Vittel unveiled the first plastic bottle aimed at a mainstream consumer market and the race was on. What started as an upscale consumer product became one of the fastest-growing industries in the world. In the early 1970s, about one billion liters of water were sold annually around the world. By 2006, global consumption had risen to close to 200 billion liters, and, with annual growth rates of around 10 percent, there is no end in sight for this sector.

Americans consume the most bottled water (32 billion liters a year), followed by Mexico (20 billion liters), China and Brazil (14 billion liters each), then Italy and Germany (12 billion liters each). Bottled water consumption is growing fastest in developing countries, particularly India (where consumption tripled in the years 2000–05), China, Mexico and South Africa (growing at 25 percent per year). Because bottled water costs anywhere from 240 to 10,000 times more than tap water, depending on the brand, the profits are very high in this sector. (For the price of one bottle of Evian, the average North American could buy 4,000 liters of tap water.) The bottled water industry is conservatively estimated to be worth us$100 billion annually.

Four companies dominate the bottled water business. Swiss food giant Nestlé started buying up successful brands, such as Vittel, Perrier, San Pellegrino and Poland Springs, in the 1990s and in 1998 launched its own water division, first called Nestlé Pure Life, now called Nestlé Waters. Company profits rose 14 percent in 2006 with bottled water accounting for almost 10 percent of company sales. Nestlé's tactic is to buy up other brands once they have become successful. With 70 different well-known brands sold in 130 countries, Nestlé is the undisputed king of the industry. Danone is Nestlé's European rival, with brands such as Evian and Volvic, and annual sales of nearly 20 billion liters – 70 percent in emerging markets – and annual profit growth of 10 percent. Both of these companies sell water from springs and underground sources and are constantly searching for new supplies around the world.

PepsiCo and Coca-Cola are the American competitors, with their Aquafina and Dasani lines in the United States and dozens of other brands internationally. Unlike their European competitors, Pepsi and Coke use tap water, which they put through reverse osmosis and to which they add minerals. Their rivalry is intense; for the first time in their 108-year history, in 2006, Pepsi beat Coke in overall sales, reports the February 2006 *Fortune* magazine even though Coke still sells way more pop than its rival. The reason: Pepsi knew before Coke that bottled water would be a hugely profitable investment and put resources into its water division that Coke is just starting to follow now, making Aquafina the number-one water brand. Coke's overall profits tumbled in 2006 by 20 percent; its only saving grace was its bottled water division, whose profits rose 10 percent. To compete, Coke has bought out Danone's stake in the bottled water industry in the United States and made a huge leap into the vitamin water industry in May 2007 with the US$4.1 billion purchase of the Whitestone, New York, company Energy Brands.

"This is an industry that takes a free liquid that falls from the sky and sells it for as much as four times what we pay for gas," Richard Wilk, a professor of anthropology at Indiana State University and who has studied the industry closely, told the *San Francisco Chronicle*. In most industrialized countries, tap water is at least as safe and often safer than bottled water, he points out. But the industry has been tremendously successful in playing on myths of declining public water systems and the few incidents of contamination that have happened to sell their product as the only safe source of water. (In early 2007, Groupe Neptune, a niche bottled water company in Paris, caused a stir when it plastered the city with fourteen hundred billboards showing a photo of an open toilet juxtaposed with a bottle of one of its brands, Cristaline, and the message, "I don't drink the water I use to flush.")

In the developing world, bottled water serves the elite; the vast majority of people cannot afford to pay for bottled water and must rely on often-polluted sources for their daily needs. It is a terrible irony that, in their competitive search for new markets, the companies often take water from poor communities in the global South to sell to upscale markets in the rich North.

Some companies and brands underscore the purity of their "product." Fiji Water boasts that it comes from an "artesian aquifer located at the edge of a primitive rainforest," whose production method guarantees that it is "untouched by man." Koyo bottling company takes a million liters of seawater from the ocean three thousand feet below the surface of the Big Island of Hawaii every day, which it then puts through reverse osmosis to be sold under the brand name MaHaLo Hawaii Deep Sea as the "purest water on earth." King Island Cloud Juice is bottled rainwater collected off the coast of Tasmania. At us$80 for a case of twelve, the character of this water changes with serving temperature, according to the company, "from sweetly refreshing when chilled to an elegant velvet when served at room temperature."

Others, such as Donald Trump's Trump Ice, which is served in his casinos and hotels, cater to an upscale market. The Source Glacier Beverage Company's 10 Thousand BC is advertised as "ultra-premium" water derived from an environmentally protected glacier in British Columbia and bottled and corked to the sound of "inspirational music." The company says its product is the "Ferrari of waters." Bling h2o is the brainchild of a Hollywood producer who noticed that on movie sets, "image is of the utmost importance . . . you could tell a lot about a person by the bottled water they carry. . . . Our product is strategically positioned to target the expanding super-luxury consumer market. It's couture water that makes an announcement like a Rolls-Royce Phantom." Bling's bottles are covered in Swarovski crystals and sell for between US$40 and US$75 for a liter bottle.

Targeting Kids

The newest target audience for the bottled water industry is children. Under pressure from parents and health advocates to remove sugar-based drinks from schools, companies have to compete with one another for bottled water brand loyalty, reports *Brandweek*. Nestlé is marketing Aquapod, shaped like a rocket and aimed at the six- to twelve-year-old market. Aquapod's slogan is "A blast of fun." Ads for the product have hit DC Comic books, Nickelodeon and children's television shows. Nestlé hired Boston-based "word-of-mouth" marketer BzzAgent to get samples to ten thousand moms and spent more than US$20 million in 2004–2005 on media. One ad depicts an old man droning on as a Bart Simpson–like kid looks at him, bored. A sign then pops down that reads, "Pull here for a blast of fun." The kid does and the old man is squashed by the Aquapod bottle.

Kids Only LLC, a leading U.S. manufacturer of children's products, has launched its own bottled water brand called Kids Only™ Bottled Water. Kids Only has teamed up with Scooby-Doo, Bratz, Superman and Spider-Man, who will adorn the bottles so that "kids can get the hydration they need in a fun way – by drinking water that comes in collectible bottles that are adorned with their favorite characters." Not to be outdone, Cott Beverages has teamed up with Disney to produce *Finding Nemo* purified drinking water and *The Incredibles* fortified flavor water. A six-pack costs US$3.99. And these are just the start. Advanced H2O has jumped into the business with Crayola Color Coolerz, while smaller companies are jumping into the market with bottled water products such as Wild Waters and WaddaJuice.

Private Companies Cash In on Blue Gold

Surging Markets

There is a very good reason that so many companies are getting in on the water business. As the world's freshwater supply dwindles, the need to find new sources rises, creating a brand-new market in a sector where there was none before. The brand-new market has created a brand-new investment opportunity, and suddenly water has become a hot property on the stock market. There are at least a dozen major water indexes as well as new exchange-traded funds dealing exclusively in water. As water analysts note, water is hot not only because of the growing need for clean water but because demand is never affected by inflation, recession, interest rates or changing tastes. "Water is a growth driver for as long and far as the eye can see," Deane Dray, water adviser for Goldman Sachs, told the *New York Times*. Lehman Brothers predicts that the number of people served globally by investor-owned water companies is expected to rise 500 percent in the next decade.

John Dickerson, of the San Diego–based Summit Water Equity Fund, says his hedge fund "universe" is comprised of 359 companies worth us$661 billion. The water business is the fastest growing of the "big three" asset industries (the others being oil and gas and electricity), says Dickerson in his 2006 annual report, with more "wind in its sails" than any other global industry. "The global theme of water insufficiency relative to unrelenting demand, along with all the related trends and opportunities it has spawned, continues to benefit the prospects of a broad range of publicly traded companies which help to provide solutions to the supply/demand dilemma. A diverse global universe of investment opportunities exists within the theme of water investing, all the while eliminating the typical resistance to sector fund investments, which are far more limited in scope and susceptible to cyclical influences. Water investing is a broad and deep global theme, and far too diverse to be considered sector investing."

Investment banking firm Seidler Capital is so enthused about prospects in the water industry, it has launched a Water Group and holds conferences called Profiting from Water – Business and Investment Opportunities in Water. "The water industry is the largest and perhaps most dynamic industry in the world," reads the brochure of one such conference held in November 2005 at the Ritz-Carlton hotel in Marina del Rey, California. "The purpose of the conference is to present a better understanding of how to achieve success in the water industry and how you can fit into the water revolution of the decade."

"Water commodities used to be thought of as boring, defensive stocks," said *MoneyWeek*, the influential British financial online journal. "Not anymore. The U.S. water sector has returned 244% over the past five years, outperforming the S&P [Standard & Poors] 500 by about 260%. . . . In the U.K., water-related utilities have also delivered some impressive returns." *Global Water Intelligence* reports that water stocks dramatically outperformed the stock markets during 2006, with their own market index up 40 percent, compared with the MSCI World Index, which improved by a modest 7.4 percent in 2006. Swiss Pictet's Water Fund, the first on the market when it was created in January 2000, advanced 22.8 percent, and Sustainable Asset Management's (SAM) Sustainability Water Fund was 20.7 percent ahead. Pictet promotes funds in forty countries with the big utility companies such as Suez and Veolia, and bottled water company Danone. SAM's fund, also Swiss, invests in such diverse companies as toilet manufacturers and water purifiers.

The French bank Société Générale created an index-certificate for the specialized World Water Index WOWAX in February 2006 and promotes investment in the twenty largest global water companies active in water provision, water infrastructure and water purification. Water is "blue gold" says the index site, with "significant" growth potential. "With the new Index Turbo warrant in the World Water Index, investors have the opportunity to profit from the development of the 20 largest global organizations in the areas of water utilities, water infrastructure and water treatment." *Bloomberg News* reports its water index of eleven utilities has returned 35 percent annually since 2003, compared with 29 percent for oil and gas stocks. In fact, says *Global Water Intelligence*, the demand for water stocks has risen faster than the supply of investment opportunities. In May 2007, Credit Suisse in partnership with Macquarie Equities launched the PL100 World Water Trust, an Australian investment fund to promote the international water industry. "The water industry resembles the oil industry during its golden era," said a director of the fund at its launch.

Investment Opportunities

There are no shortages of water investment opportunities for those with money. Jim McWhinney of Investopedia, an Alberta, Canada–based online investment advice service, explains that "like any other scarcity, the water shortage creates investment opportunities and interest in water is at an all-time high." Shares in ITT climbed 135 percent between 2002 and 2007. U.S.–based Pentair, a tool company, bought the water treatment company Wicor in 2004 and with sister Met-Pro subsidiary, Pristine Water Solutions, is raking in the money. Company president Raymond De Hont told the *New York Times*, "Ten years ago, we were a US$100 million pump company. We're now doing US$2.13 billion in water alone." Other recent U.S. players cashing in on the water bonanza include WaterBank of America, a global supplier of ice for luxury hotels and cruises created in 2002, which has acquired a "bank" of top-quality freshwater springs to supply its business; Watts Water Technologies, a supplier of industrial water technology parts with seventy-five facilities worldwide; and Itron, which manufactures and operates water meters around the world.

High on the list of indexes to watch is Palisades Water Index, a group of thirty-seven water companies, five of them foreign, which operates in the U.S. market. This index is designed to track the performance of companies involved in the global water industry and has climbed an average of 18.7 percent every year since 2001. The American Stock Exchange (AMEX) began publishing the Palisades Water Index over the Consolidated Tape on December 11, 2006. The index is now published daily every fifteen seconds. A new exchange-traded fund, the PowerShares Water Resources Portfolio, set up in late 2005 to track the performance of the Palisades Water Index, has already attracted about US$1 billion in assets – four times the average haul of other exchange-traded funds launched at the same time, reports Tim Middleton of MSN Money, an online market reporting service.

The Media General Water Utilities Index is up 133 percent over the decade, and Paris-based Praetor Global Water Fund is showing similar growth. New York–based Terrapin's Water Fund returned 22 percent in the first year after it was started in April 2005. The Dow Jones U.S. Water Index, composed of twenty-three stocks that grew a whopping 221 percent between 2000 and 2006, is another major player as are ISE-B&S Water Index, launched in 2006 by the International Securities Exchange "in response to growing interest in water as a scarce resource," and S&P 1500 Water Utilities Index, composed of only two companies: American States Water and Aqua America, the biggest U.S.–based private water utilities, with close to three million clients between them. The Bloomberg World Water Index and the MSCI World Water Index provide information on the global water business.

Green Cover

A number of water funds market themselves as "environmental," but their privatization and profit-making agenda is clear. The Global Environment Fund is a U.S.–based SEC registered investment fund with assets of close to US$1 billion and a mandate to invest in sustainable practices. However, the fund openly boasts about the role it played in forcing the privatization of the Brazilian state-owned water company SANEPAR: "GEF has invested nearly US$30 million in [SANEPAR], and plays an important role in the consortium of strategic investors who have taken over the management of the company."

The Global Environment Emerging Markets Fund is a US$70 million private equity fund related to GEF with investments in private water companies operating in the Third World. One investment in Africa has included a continent-wide water treatment facility with a U.S.–based consumer products company. Atlantis is another "green" private equity fund worth us$250 million, launched as a joint venture between the Global Environment Fund and Poseidon Resources to promote private desalination plants in developing countries.

Aqua International Partners is a private water equity fund founded by William K. Reilly, former head of the Environmental Protection Agency (EPA), whose mandate is to "invest in operation and special purpose companies that are devoted to: bottled water; water purification and treatment operations – either being privatized or newly formed; manufacturing equipment or products (e.g. pipes, pumps, meters, filters, et al.) for commercial, industrial and residential water users; other related activities or services supporting the purification, treatment and supply of water." During his term as head of the EPA (1989–1993), Reilly had responsibility for the largest program of finance of water supply and wastewater treatment by the U.S. government, overseeing more than us$8 billion in water investments. In his new role, he has been well placed to take advantage of millions in seed funding for Aqua International from the U.S. Overseas Private Investment Company, a U.S government agency that "supports U.S. investments in emerging markets worldwide, fostering the growth of free markets."

How big is the global water industry? Over and over, one reads that the water market is us$400 billion annually. But this is a dated statistic and usually refers only to the big service companies and the more traditional waste treatment sector. The Summit Water Equity Fund alone handles water companies worth almost us$700 billion. It is clear that if bottled water is included, as well as the billions that will go into urban water infrastructure repair in the near future, plus the emerging technologies of purification, desalination and nanotechnology, the global water market can conservatively be said to be a trillion-dollar-plus-per-year industry with no limit in sight. Add shipping companies to ship water around the world as the business of water trading and bulk exports grows, as well as pipeline and construction companies to lay the global network of pipes now in the planning stage, and the estimate soars into the trillions. Then add nuclear power to fuel the industry. Ask any stockbroker: there is no limit to the money to be made in water.

The Corporate Takeover of Water
Deepens the Global Water Crisis

It is evident that the world is moving toward a corporate-controlled freshwater cartel, with private companies, backed by governments and global institutions, making fundamental decisions about who has access to water and under what conditions. It is unlikely that there will come a time when there is no private involvement in water. Nor are most critics saying there is no place for private companies in finding solutions to the coming global water crisis. However, there is a desperate need for public oversight and control of the world's declining water supply and for elected governments, not corporations, to make the decisions about this shared heritage before it is too late.

One unanswered question, for example, is who will own the water recycled by water reuse corporations. Arguably, a private company could claim that it has purified dirty water and now owns the finished "product." Certainly, the bottled water companies currently own the water they treat and sell. Imagine a day when the majority of our water comes to us recycled by private companies: Will they own the actual water or just the right to make a profit from having cleaned it? If they own it, will they be able to decide who lives and dies?

Simply put, the answer to the world's water crisis rests on the principles of conservation, water justice and democracy. No global corporation that must be competitive to survive can operate on these three principles. There are three major problems with the growing corporate control of water.

There Is No Incentive to Stop Pollution

The first is that there is no profit in conservation. In fact, it is to the distinct advantage of the private water industry that the world's freshwater supplies are being polluted and destroyed. Even if individual corporate leaders do not take pleasure in the global water crisis, it is exactly this crisis that is driving profits in their industry. The "dead hand" of the market will favor those companies that maximize profit and, in the water business, that means taking advantage of a dwindling supply that cannot meet a growing demand. (In the minutes of several November 2005 confidential meetings of RWE's American Water, obtained by Food and Water Watch, CEO Harry Roels bemoaned the fact that extra costs incurred to the company from environmental regulation could not be passed on to the consumers, as they responded to rate hikes by lowering demand. If conservation was the goal of the company, such a reaction would be a good thing.) Further, with governments, industries and universities investing so heavily in the burgeoning water cleanup technology industry, there is less and less incentive at every level to emphasize source protection and conservation. Once a massive and expensive cleanup industry is in place, economic and political pressure will come to bear on governments and global institutions to protect it. Technology, controlled by corporations, will drive policy.

Already global trade rules to promote the water technology industry are in place. The World Trade Organization promotes and protects the trade in environmental services, encouraging cross-border trade and investment in private water cleanup companies. As in all tradable goods and services, governments are encouraged to relinquish public control of water treatment to the private sector and have to ensure that any rules they have in place are the least trade-restrictive possible. This means that their rules and regulations meant to protect the public and the environment must not hamper private business, and the pressure is on governments to "cut red tape" and lower their standards. As well, under the National Treatment provision of the WTO, governments cannot favor domestic water companies and will have to open up their bidding process to the water technology transnationals that are getting more powerful all the time.

Only the Rich Will Have Clean Water

The second problem with corporate control of water is that water and water infrastructure – from drinking water and sanitation utilities services to bottled water, cleanup technologies and nuclear-powered desalination plants – will flow where the money is, not where it is needed. No corporation is in business to deliver water to the poor. That, say corporate leaders, is the job of governments. People who cannot pay do not get served.

Already, wealthy countries such as Saudi Arabia and Israel are dependent on expensive water purification technologies for their day-to-day living, while equally water-starved countries such as Namibia and Pakistan cannot afford such technology, and so their citizens suffer from severe water shortages. Bottled water is the exclusive prerogative of those who can pay for it, as is clean water from the tap in many parts of the world. World Water and Flow Inc., the two companies on the verge of a bulk water transfer business, are looking to send their first shipments not to the parts of the world where people are dying for water but to Las Vegas and Los Angeles in the case of World Water and Saudi Arabia and the United Arab Emirates in the case of Flow.

Further, as in every major industry sector, the water industry is becoming very powerful in lobbying and advising governments and global institutions on water policy. As outlined in Chapter 2, the big service companies have enormous clout with the World Bank and the United Nations as well as with their own governments. *Pipe Dreams* reports that the big companies such as Suez and Veolia actually influence World Bank decisions as to where funding for water services should go. "Putting private companies in the driving seat in recent years has allowed them to set the agenda in terms of prioritising the contents, regions and cities where investment in the water sector should go." Because of the corporate need to make a profit, donor-funded investments have not concentrated on the areas of greatest need, be it by country or by city where the greatest number of poor live. Rural communities have suffered as well from lack of attention because of their inability to create a profit for the water companies. As a consequence, sub-Saharan Africa and South Asia have been the focus of only 1 percent of total promised private sector water investment.

Nature Will Have to Fend for Itself

The third major concern around corporate control of water is that, with no regulatory oversight or government control, there will be no protections for the natural world and the need to safeguard integrated ecosystems from water plundering. As it is now, in most parts of the world, governments have little knowledge of where their groundwater sources are located or how much water they contain; consequently, they have no idea how much pumping they can maintain or if current water mining operations are sustainable. The more private interests control water supplies, the less government and public interests have to say about them. The commodification of water is really the commodification of nature. If water in the future will only be accessible to those who can pay for it, who will buy it for nature?

An added strain is put on rural and wilderness water sources by the water needs of urban centers, especially the burgeoning megacities of the developing world, needs increasingly being supplied by draining rural and wilderness lakes, rivers and aquifers. If governments maintain control of water systems, they can try to protect rural ecosystems, although it is true that governments are under competing pressures. But if, as is increasingly the case, water transfers are in the hands of private brokers who are competing with one another for dwindling resources and the process is unregulated by governments, there will be few protections in place to stop the destruction of watersheds and ecosystems and the species and plant life they sustain.

Further, every one of the new water technologies, including water purification, recycling and nanotechnology, and many of the current water practices, including bottling water and pipeline transportation of water, in and of themselves pose a direct risk both to the natural world and to human health.

Water Purification and Recycling

The environmental problems of desalination were covered in Chapter 1. They include the fact that desalination plants are energy-intensive; they may use seawater in the intake process that is polluted with waste that has been dumped untreated into the ocean; and they produce a lethal by-product. In fact, the production of a lethal by-product is standard for all water purification systems, as they need to use chemicals to protect their osmosis membranes and must expel these chemicals together with the unwanted ingredients and contaminants that have been removed during the purification procedure. Purification processes are also energy-intensive.

Increasingly, there is interest in using purification systems to recycle wastewater for direct reuse, including for cooking and drinking, with "experts" assuring governments that recycled water is as safe as spring water. While few will argue with the need to use recycled water for industrial use in the workplace or gardens, toilets and house cleaning in the home, there are real concerns about bathing, drinking or cooking with recycled water. Recent studies reveal that treated water can contain residues from a whole host of toxic substances, including pharmaceuticals, hormones, antibiotics, chemotherapy medication, birth control and endocrine-disrupters – chemicals that mimic the effects of estrogen and have been linked to sexual abnormalities and disorders in animals and humans. UCLA and University of Wisconsin scientists Mel Suffet, Joel Pedersen and Mary Soliman report that even the most advanced ultra-filtration purification technologies will not remove these and other contaminants. They studied the effluent from three high-tech effluent recycling plants in Los Angeles for fifty-four chemicals, pharmaceuticals, hormones and toxins that are not part of the regulated monitoring protocol. Each of the plants had between twenty-nine and thirty-four of these toxins in the highly treated effluent.

U.S. cancer expert Dr. Steven Oppenheimer, director of the Center for Cancer and Development Biology at California State University, Northbridge, has raised similar concerns. Oppenheimer told the *West Australian* that the "toilet to tap" process should only be considered as a last resort, likening drinking recycled water to playing "Russian roulette" with human life. "The world's scientific community does not and will not know all the toxic agents and carcinogens that may be able to make it through the indirect reclaimed water process to drinking water," he said. Surveys by the American Water Works Association in Philadelphia and New Jersey have found traces of drugs, herbicides, fragrances, hormones and weed killers in their treated water. Christopher Crockett, manager of watershed protection at the Philadelphia Water Department, is concerned. "In our preliminary study, we found all the compounds the New Jersey study found, including endocrine disrupters," he told the *Philadelphia Inquirer*. The May 2007 edition of the international journal *Water Research* reported that small concentrations of antibiotics have the ability to pass through even advanced wastewater treatment plants. The study was carried out by the National Research Center for Environmental Toxicology at the University of Queensland in Australia. As well, a June 2007 issue of *Newsweek* sounded the warning that the effects of endocrine disrupters "add up" in the human body and cites many studies linking feminized male fish living downstream from sophisticated First World treatment plants. The article quotes scientists wondering if these toxins could be the cause of other well-documented trends, including early puberty in girls.

At the very least, this situation calls for strict government control and testing, as some countries are actively moving toward water reuse. And no recycled water should be permitted for drinking until it is proven completely safe by independent testing.

However, the industry is pressuring governments to approve recycled water for all uses because it is too expensive for people to install two pipe systems into their homes – one for drinking and the other for recycled water. In *Water Reuse Markets 2005–2015: A Global Assessment & Forecast*, a special report published by Global Water Intelligence, the authors bemoan the fact that the main drawback of water reuse is that it is not considered suitable for drinking and cooking because water reuse would need a costly and separate new distribution infrastructure, which would in turn increase the capital costs involved in the projects.

The answer? "A change in policy to enable direct potable reuse would reduce the operating cost of new water reuse projects by 30%, greatly enhancing the scope of the market." The main beneficiaries? "Membrane manufacturers and process engineers." Little wonder that the industry is seeking the deregulation of the water market. "The water industry is where the telecommunications industry was 20 years ago, highly regulated and on the verge of a major change," Ori Yogev, chairman of Waterfront, a newly formed Israeli water lobby, told *BusinessWeek*. In its March 2007 report, *Investing in Water*, Progressive Investor, a leading sustainable investment research firm, identified public control of water as a "barrier to growth." With government oversight gone, the sky's the limit in terms of reuse and other technologies.

Regulating Nanotechnology

Similar resistance to government oversight is found in the emerging water nanotechnology industry, which is being forced to counter the alarm bells going off within some scientific and environmental communities over free nanoparticles – mobile particles able to separate from the property into which they were introduced and roam freely in nature or the body. There, they can find their way into skin, lungs, livers and kidneys and even break the blood-brain barrier.

The situation is related to the size of the technology: the smaller the particle, the larger its surface in relation to its volume. Nanoparticles have an enormous surface-to-volume ratio, which renders them biologically active. Mark Wiesner, chemical engineering professor at Rice University in Texas, has found that nanoparticles do not flow uniformly in water, and he is calling for a slow-down on development of the technology until more independent research can be done. Wiesner's research, presented at the 2004 annual meeting of the American Chemical Society, has found that the ways in which these particles behave in groundwater environments or water treatment plants are as varied as the diverse molecules or atoms used to assemble them. "When you get smaller, properties change," Wiesner told the *Associated Press*. Other scientists at the conference reported that certain nanoparticles caused brain damage in fish.

Several major U.S. environmental and health organizations are demanding strict government oversight of this emerging technology. In May 2006, Friends of the Earth, Greenpeace and the International Center for Technology Assessment called for the U.S. Food and Drug Administration to treat nanoparticles as "new substances" and subject them to rigorous health and safety testing before being allowed onto the market. They join the British Royal Society, who note that nanoparticles are different from anything humans have ever created before; the Royal Society has called for Great Britain to adopt the precautionary principle (that a substance must be proven safe before it is allowed on the market) in its approach to this technology. "Until there is evidence to the contrary," says the society, "factories and research laboratories should treat manufactured nanoparticles and nanotubes as if they were hazardous and seek to reduce them as far as possible from waste streams."

However, just the opposite of government control is taking place. Critics warn that a whole new world of ownership and corporate control has been opened up with this technology and that, just as biotechnology led to corporate patenting of life, nanotechnology is going to lead to corporate patenting of matter unless it is stopped. The Natural Resources Defense Council cautions on its website that without government and public control, "we will be allowing the nanotechnology industry to conduct an uncontrolled experiment on the American people." So far, nanoparticles are not subject to any special government regulation although there is growing pressure for governments to act. And that's just fine with the industry. At the first meeting of the U.S. Food and Drug Administration Nanotechnology Task Force held in October 2006 in Bethesda, Maryland, representatives of the industry insisted that existing rules are adequate. "The FDA already has in place a comprehensive regulatory system," Matthew P. Jaffe, an attorney speaking on behalf of the United States Council for International Business told *The Scientist*, while admitting that there are no FDA regulations that specifically address nanomaterials.

Bottled Water

The bottled water industry is one of the most polluting industries on Earth, and one of the least regulated. Most bottled water comes in plastic made of polyethylene terephthalate (PET), derived from crude oil, and chemicals, such as polyethylene and phthalates, which can leach from the bottle into the water and certainly into the ground. Nearly one-quarter of all bottled water crosses national borders to reach consumers, using enormous amounts of energy to fuel the boats, trains and trucks carrying it. One million bottles of exported drinking water causes the emission of 18.2 tons (18,000 kilograms) of carbon dioxide. Worldwide, 2.7 million tons (about 2.5 billion kilograms) of plastic are used to bottle water every year, creating mountains of garbage and fouling waterways. Fewer than 5 percent of plastic bottles around the world are recycled; most are either incinerated, which produces toxic by-products such as chlorine gas and ash containing heavy metals, or buried, where they take a thousand years to biodegrade. Of those plastic water bottles from the global North that are recycled, almost half go to China for processing, where they are fouling China's already-strained waterways and adding to the energy costs of this product.

As well, water extraction for bottled water is concentrated near already-stressed water systems such as the Great Lakes (from which almost four trillion liters of water are already taken from the basin every day, according to the *Detroit News*) or in rural communities, such as the many Coca-Cola plants in rural India, where the sudden drop in water affects fewer people but devastates the watersheds and livelihoods of those who live there. Water bottlers pay almost nothing for the water they extract and in most countries pay little or no royalties or taxes on this common heritage from which they earn such huge profits. Further, large amounts of water are wasted in the production of Pepsi and Coke bottled water, which is essentially filtered tap water; it takes 2.6 liters of tap water to produce 1 liter of their bottled water due to the complicated and wasteful filtering process. (In addition, it takes 250 liters of water to produce the sugar for any of its flavored waters or soft drinks.)

Nor is bottled water any safer than tap water on the whole. Studies have found that because it is largely unregulated, some bottled water is, in fact, less safe than more highly regulated tap water. Among these studies are the well-known 1999 ground-breaking survey by the U.S.–based Natural Resources Defense Council, which took four years to produce and examined 1,000 different bottles and 103 brands; a 2004 study of 68 brands of European mineral water by Dr. Rocus Klont of the University Medical Center in Netherlands, which found "high levels of bacterial contamination" including traces of legionella bacteria as well as mold penicillium, in the bottled water; and a 2006 report called *Have You Bottled It?* by the British environmental group Sustain, which recommends tap water over bottled water for personal health and the planet. Coca-Cola was forced to recall all its Dasani bottled water from the United Kingdom in 2004 when it was discovered to have high levels of bromate, a chemical compound that can cause abdominal pain, hearing impairment, kidney failure and even death at high-enough doses.

Yet companies such as Coca-Cola are aggressively marketing their water as some kind of miracle beverage in schools and universities around the world and finding new markets as people buy the myth that only bottled water is safe to drink. Perhaps the worst thing about bottled water is that it allows people to view water as a commodity and sets the stage – one bottle at a time – for acceptance of the complete corporate takeover of water.

⁓⁓⁓⁓

Ironically, the explosion of the bottled water industry has also created a powerful backlash against the commodification of water that forms one important pillar of the global water justice movement. The planned private takeover of water services in the developing world, combined with the sudden emergence of new water technologies and the creation of a global corporate water cartel, was bound to create a grassroots backlash from citizens and communities around the world. For the last decade, a global citizens movement has challenged the growing political influence of transnational corporations in every sphere of life as well as the unsustainability of unlimited growth. In particular, activists have fought the privatization of "the commons," those areas of life once thought to be the common heritage of humanity for the benefit of the many, now coming under corporate control for the benefit of the few. With scarce water already a source of contention, it is hard to imagine that the water companies, the World Bank and the political actors who support them did not see the backlash coming.

Chapter 4

The Water Warriors Fight Back

> *Thousands have lived without love, not one without water.*
>
> – W. H. Auden, *First Things First*

A fierce resistance to the corporate takeover of water has grown in every corner of the globe, giving rise to a coordinated and, given the powers it is up against, surprisingly successful water justice movement. "Water for all" is the rallying cry of local groups fighting for access to clean water and the life, health and dignity that it brings. Many of these groups have lived under years of abuse, poverty and hunger. Many have been left without public education and health programs when their governments were forced to abandon them under World Bank structural adjustment policies. But somehow, the assault on water has been the great standpoint for millions. Without water there is no life, and for thousands of communities around the world, the struggle over the right to their own local water sources has become a politically galvanizing milestone.

A mighty contest has grown between those (usually powerful) forces and institutions that see water as a commodity, to be put on the open market and sold to the highest bidder, and those who see water as a public trust, a common heritage of people and nature and a fundamental human right. The origins of this movement, generally referred to as the global water justice movement, lie in the hundreds of communities around the world where people are fighting to protect their local water supplies from pollution, destruction by dams and theft – be it from other countries, their own governments or private corporations such as bottled water companies and private utilities backed by the World Bank.

Until the late 1990s, however, most were operating in isolation, unaware of other struggles or the global nature of the water crisis.

Latin America

Latin America was the site of the first experimentation with water privatization in the developing world. The failure of these projects has been a major factor in the rejection of the neo-liberal market model by so many Latin American countries who have said no to the extension of the North American Free Trade Agreement to the southern hemisphere and who have forced the big water companies to retreat. A number of Latin American countries are also opting out some of the most egregious global institutions. In May 2007, Bolivia, Venezuela and Nicaragua announced their decision to withdraw from the World Bank's International Centre for the Settlement of Investment Disputes, in no small measure because of the way the big water corporations have used the center to sue for compensation when the countries terminated private delivery contracts.

Latin America, with its water abundance, should have one of the highest per capita allocations of water in the world. Instead, it has one of the lowest. There are three reasons, all connected: polluted surface waters; deep class inequities; and water privatization. In many parts of Latin America, only the rich can buy clean water. So it is not surprising that some of the most intense fights against corporate control of water have come out of this region of the world.

Bolivia

The first "water war" gained international attention when the indigenous peoples of Cochabamba, Bolivia, led by a five-foot, slightly built, unassuming shoemaker named Oscar Olivera, rose up against the privatization of their water services. In 1998, under World Bank supervision, the Bolivian government had passed a law privatizing Cochabamba's water system and gave the contract to U.S. engineering giant Bechtel, which immediately tripled the price of water, cutting off those who could not pay. In a country where the minimum wage is less than US$60 a month, many users received water bills of US$20 a month, which they simply could not afford. The company even charged them for rainwater they collected in cisterns. As a result, La Coordinadora de Defensa del Agua y de la Vida (the Coalition in Defence of Water and Life), one of the first coalitions against water privatization in the world, was formed and organized a successful referendum demanding the government cancel its contract with Bechtel. When the government refused to listen, many thousands took to the streets in non-violent protest and were met with army violence that wounded dozens and killed a seventeen-year-old boy. On April 10, 2000, the Bolivian government backed down and told Bechtel to leave the country.

The Bolivian government had also bowed to pressure from the World Bank to privatize the water of La Paz and in 1997 gave Suez a thirty-year contract to supply water services to it and neighboring El Alto, the hilly region surrounding the capital where thousands of indigenous peoples live. From the beginning, there were problems. Aguas del Illimani, the Suez subsidiary, broke three key promises: it did not deliver to all the residents, poor as well as rich, leaving about two hundred thousand without water; it charged exorbitant rates for water hookups, about US$450, equivalent to the budget of a poor family for two years; and it did not invest in infrastructure repair and wastewater treatment, choosing instead to build a series of ditches and canals through poor areas of La Paz, which it used to send garbage, raw sewage and even the effluent from the city's abattoirs into Lake Titicaca, once prized as a UNESCO World Heritage site. To add insult to injury, the company located its fortresslike plant under the beautiful Mount Illimani, where it captured the snowmelt off the mountain, and, after rudimentary treatment, sent it by pipe into the homes of families and businesses in La Paz who could pay. The nearest community, Solidaridad, a slum of about one hundred families with no electricity, heat or running water, had its only water supply cut off. Its school and health clinic, built with foreign-aid money, could not operate because of a lack of water. It was the same all through El Alto.

An intense resistance to Suez formed. FEJUVE, a network of local community councils and activists, led a series of strikes in January 2005, which crippled the cities and brought business to a halt. This resistance was a prime factor in the ousting of presidents Gonzalo Sanchez de Lozada and Carlos Mesa. Their replacement, Evo Morales, the first indigenous president in the country's history, negotiated Suez's departure. On January 3, 2007, he held a ceremony at the presidential palace celebrating the return of the water of La Paz and El Alto after a long and bitter confrontation. "Water cannot be turned over to private business," said Morales. "It must remain a basic service, with participation of the state, so that water can be provided almost for free."

Argentina

Rio de la Plata (the Silver River) separates Buenos Aires, the capital of Argentina, from Montevideo, the capital of Uruguay. For five hundred years, it has also been called la Mar Dulce (the Soft Sea) because its size made people think it was a freshwater sea. Today, however, the river is famous for something else: it is one of the few rivers in the world whose pollution can be seen from space. On March 21, 2006, the Argentine government rescinded the thirty-year contract of Aguas Argentinas, the Suez subsidiary that had run the Buenos Aires water system since 1993, in no small part because the company broke its promise to treat wastewater, continuing to dump nearly 90 percent of the city's sewage into the river. As well, in another broken promise, the company repeatedly raised tariffs, for a total increase of 88 percent in the first ten years of operation. Water quality was another issue; water in seven districts had nitrate levels so high, it was unfit for human consumption. An April 2007 report by the city's ombudsman stated that most of the population of 150,000 in the southern district of the city lived with open-air sewers and contaminated drinking water.

Yet, as Food and Water Watch reports, the Inter-American Development Bank continued to fund Suez as late as 1999, in spite of the mounting evidence that the company was keeping its large profits, pulling in 20 percent profit margins while refusing to invest in services or infrastructure. Outrageously, with the backing of the French government, Suez is trying to recoup US$1.7 billion in "investments" and up to US$32 million in unpaid water bills at the World Bank arbitration court, the International Centre for the Settlement of Investment Disputes (ICSID). Suez had just (in December 2005) been forced out of the province of Santa Fe, where it had a thirty-year contract to run the water systems of thirteen cities. The company is also suing the provincial government at the ICSID for US$180 million. Close on the heels of the Buenos Aires announcement, Suez was forced to abandon its last stronghold in Argentina, the city of Cordoba, when water rates were raised 500 percent on one bill.

In all cases, strong civil society resistance was key to these retreats. A coalition of water users and residents of Santa Fe, led by Roberto Munoz and others, actually organized a huge and successful plebiscite, in which 256,000 people, more than one-quarter of the population of the province, voted to rescind Suez's contract. They convened a Provincial Assembly on the Right to Water with 7,000 activists and citizens in November 2002, which set the stage for the political opposition to the company. The People's Commission for the Recovery of Water in Cordoba is a highly organized network of trade unions, neighborhood centers, social organizations and politicians with a clear goal of public water for all, and was instrumental in getting the government to break its contract with Suez. "What we want is a public company managed by workers, consumers and the provincial government, and monitored by university experts to guarantee water quality and prevent corruption," says Luis Bazán, the group's leader and a water worker who refused employment with Suez.

Mexico

Mexico is a beachhead for privatization across the region, with its elites having access to all the water they need and also controlling governments at most levels of the country. Only 9 percent of the country's surface water is fit for drinking, and its aquifers are being drawn down mercilessly. According to the National Commission on Water, twelve million Mexicans have no access to potable water whatsoever and another twenty-five million live in villages and cities where the taps run as little as a few hours a week. Eighty-two percent of wastewater goes untreated. Mexico City itself has dried up, and its twenty-two million inhabitants live on the verge of crisis. Services are so poor in the slums and out-skirts of the city, cockroaches run out when the tap is turned on. In many "colonias" in Mexico City and around the country, the only available water is sold from trucks that bring water in once a week, often by political parties that sell the water for votes.

In 1983, the federal government handed over responsibility for the water supply to the municipalities. Then in 1992, it passed a new national water bill that encouraged the municipalities to privatize water in order to receive funding. Privatization was supported by former president Vicente Fox, himself a former senior executive with Coca-Cola, and is also favored by the current president, Felipe Calderon. The World Bank and the Inter-American Development Bank are actively promoting water privatization in Mexico. In 2002, the World Bank provided us$250 million for infrastructure repair with conditions that municipalities negotiate public-private partnerships. Suez is deeply entrenched in Mexico, running the water services for Mexico City, Cancun and about a dozen other cities. Its wastewater division, Degremont, has a large contract for San Luis Potosi and several other cities as well. The privatization of water has become a top priority for the Mexican water commission, CONAGUA. As in other countries, privatization in Mexico has brought exorbitant water rates, broken promises and cutoffs to those who cannot pay. The Water Users Association in Saltillo, where a consortium of Suez and the Spanish company, Aguas de Barcelona, run the city's water systems, reports that a 2004 audit by the state comptroller found evidence of contractual and state law violations.

A vibrant civil society movement has recently come together to fight for the right to clean water and resist the trend to corporate control. In April 2005, the Mexican Center for Social Analysis, Information and Training (CASIFOP), brought together more than four hundred activists, indigenous peoples, small farmers and students to launch a coordinated grassroots resistance to water privatization. The Coalition of Mexican Organizations for the Right to Water (COMDA) is a very large coalition of environmental, human rights, indigenous and cultural groups devoted not only to activism, but also to community-based education on water, its place in Mexico's history and the need for legislation to protect the public's right to access. Their hopes for a government supportive of their perspective were dashed when conservative candidate Felipe Calderon won (many say stole) the 2006 presidential election over progressive candidate Andrés Manuel Obrador. Calderon is working openly with the private water companies now to cement private control of the country's water supplies.

Chile
Chile has had almost total private water delivery for a decade, serviced mostly by British companies. Resistance to water privatization in Chile is very difficult because of the entrenched commitment to market ideology of the ruling elites. Neo-liberal market reform was the cornerstone of dictator Pinochet's policy and the excuse for the atrocities carried out during his regime. Chile was one of the first countries to privatize all aspects of government services from health to education and, later, electricity and water. In fact, at first, Chile adopted the British model of water privatization introduced by Margaret Thatcher in Great Britain, whereby the companies bought and controlled the whole system. But mounting concerns over the environmental impacts of having whole water systems owned by foreign companies forced the government to change direction and decree that only

the leasing and management models (where at least there is some element of public control maintained) would be considered in future contracts.

As with other water privatization schemes in Latin America, water rates in Chile have steadily climbed, putting access to adequate water supplies out of bounds for millions. As Food and Water Watch reports, even the government admits that rates have risen 20 percent; but citizens groups have documented much higher rate hikes, as high as 200 percent in some communities. That privatization is deeply ideologically entrenched in Chile's ruling class was clear when, in a 2000 plebiscite, 99.2 percent of voters in Chile's central valley region rejected water privatization but the government privatized local services anyway. There is hope that the new center-left government of Michelle Bachelet will be more open to arguments for public governance of Chile's water supplies.

Civil society groups have had more success in their campaign against another threat to Chile's water sources, namely, the threat by notorious Canadian mining company Barrick Gold to remove the top of three glaciers on the Chile-Argentine border in order to get at the gold deposits underneath them. Barrick has been given the go-ahead to take 500,000 kilograms of gold from the Pascua-Lama mine, in itself still a controversial operation. But the original plan meant removing, by means of blasting and bulldozing, 826,000 cubic meters of glacier ice at the headwater of a basin that serves as the major source of water for the 70,000 small farmers of the surrounding area. Now, instead of the open-pit mine the company envisaged, it has to blast into the mountain – a much more expensive operation. Meanwhile, environmentalists, led by former presidential candidate Sara Larrain and her group, Sustainable Chile, obtained a signed agreement from President Bachelet not only to protect the glaciers for all time, but also to create a new environment ministry that will invest in and protect the nation's natural heritage, including water.

Ecuador

On March 1, 2007, the mayor of Quito, Ecuador's capital, announced he had halted plans to privatize the city's water, plans that had been seriously underway for four years. The Coalition for the Defense of Public Water used a Price Waterhouse report to show that the city would have had to invest US$20 million in the first five years while the private companies would only have invested US$7 million to make the project work. After just six years, the company would start to make a profit, worth US$226 million over thirty years. The coalition was also able to tell the story of how the Bechtel subsidiary, Guayaquil Interagua, behaved in early 2001 when it took over the water services of Guayaquil, Ecuador's largest city, with two million inhabitants. The company immediately fired all the workers, started dumping 95 percent of its wastewater into the local rivers, leading to a major hepatitis A outbreak in 2005, and cut off water to thousands who could not afford it. Food and Water Watch reports that the local citizens network, Citizens' Observatory for Public Services, is demanding that the government fine the company for its violations.

Others

Other Latin American cities or countries rejecting water privatization include Bogotá, Colombia (although other Colombian cities, including Cartagena, have adopted private water systems), Paraguay, whose lower house rejected a Senate proposal to privatize water in July 2005; Nicaragua, where a fierce struggle has been waged by civil society groups and where in January 2007 a court ruled against the privatization of the country's wastewater infrastructure; and Brazil, where strong public opinion has held back the forces of water privatization in most cities. Unfortunately, resistance in Peru, where increased rates, corruption and debt plague the system, has not yet reversed water privatization.

Asia–Pacific

Almost every Asia-Pacific country has either introduced private management or is considering it. The World Bank and the Asian Development Bank (ADB) have actively promoted the big water companies throughout the region and in 2006 formed a regional arm of the World Water Council called the Asia-Pacific Water Forum, holding its first meeting in December 2007 in Japan. The private companies and the ADB are becoming more organized around strategy in the face of growing and intense opposition from every community where privatization has been implemented. In December 2003, a new Asia-Pacific network to protect the right to water was launched in Bangkok at a conference sponsored by Jubilee South and the Asia-Pacific Movement on Debt and Development. The campaign, called People's Right to Water and Power, committed the new network to oppose privatization of water in the Asia-Pacific, work to remove water from all trade and investment agreements of the WTO, expose the linkages between a poor country's debt and its powerlessness to oppose privatizations, and promote wider recognition and institutionalization of water as a human right. In May 2007, hundreds of members of the Freedom from Debt Coalition stormed the gates of the Asian Development Bank's fortieth annual governors' meeting in Manila to protest the ecological destruction and growing poverty imposed on the region by the banks' policies, including the promotion of private water services.

India

India has treasured a hard-won tradition of community care and control over resources. But in the last few years, with the creation of its new entrepreneurial class, India has begun to adopt the Washington Consensus model in many sectors. Water privatization was promoted in the 2002 National Water Policy, which called for private water services "wherever feasible." A year later, India's Ministry of Urban Development released guidelines to state governments blaming "unreliable flows of public funds" for the water crisis and called on them to create a "welcoming atmosphere" in the drinking-water sector. The ministry knew this would be contentious; in 2000, furious farmers in Andhra Pradesh chased then World Bank head James Wolfensohn away from a pro-privatization public event sponsored by politicians friendly to the World Bank agenda. The ministry even admitted that consumers would bear the burden of this change with tenfold increases in water rates, an announcement met with widespread criticism. India is now in the midst of a furious round of privatization. The water corporations are all over the country, vying for contracts in municipalities and buying up whole river systems.

Bechtel is supplying water and sewage services in Tamil Nadu. Veolia is operating in Jamshedpur, Agra, Calcutta and Visakhapatnam. Thames is eyeing domestic water supply in Indore. Anglian is competing for water distribution in Mysore, Mangalore, Hubli and Dharwad. Suez, operating under its subsidiary Degremont, has projects in Delhi, Bangalore, Chennai and Nagpur. Its Delhi operation is particularly contentious, as the contract for the treatment plant being developed by the company to build and operate is backed with government guarantees of profit. As well, the US$50 million project includes the construction of a giant thirty-kilometer pipeline that will divert water from the Upper Ganga Canal through the World Bank–backed Tehri Dam to supply drinking water to Delhi. Thousands have been forcibly removed from their homes and farms for this highly controversial project, which also diverts the sacred waters of the Ganges. Under intense resistance from water rights groups such as Navdanya, the Citizens Front for Water Democracy, the Indian government is furiously backpedaling, assuring the Indian people that these projects are not privatizations but public-private partnerships where public control is maintained.

The government has also (temporarily) backed away from its highly questionable plan to link fourteen Himalayan rivers and sixteen rivers in the global South in order to irrigate huge swaths of farmland, in the face of heavy criticism about potential catastrophic environmental impacts and the displacement of millions. Less successful have been the fights to stop the sale of whole rivers to private corporations, such as the Sheonath River in Chhattisgarh, where a private consortium has a twenty-two-year lease for the exclusive use of a twenty-seven-kilometer river. India also has a fierce anti-bottled water movement, as companies such as Coca-Cola and PepsiCo. have set up plants all over the rural countryside, mining precious and dwindling water sources, and creating great distress.

India is also the home to one of the most powerful anti-dam movements in the world, with high-profile leaders such as Medha Patkar, Vandana Shiva and Arundhati Roy. The most intense struggle, led by the tenacious grassroots movement Narmada Bachao Andolan, is the fight to stop the Sardar Sarovar dam, the largest of thirty large dams, and more than three thousand small and medium-sized dams planned for the mighty Narmada and its tributaries, which will displace close to one million people – mostly tribal villagers and small farmers – from their land.

Indonesia

With the blessing of and funding from the World Bank and the Asian Development Bank, Suez and Thames Water used their connections to the regime of former Indonesian dictator Suharto to secure concessions to Jakarta's water, which was privatized in 1998, without public consultation or bidding. It is well documented that the companies reneged on their signed contracts to improve water delivery to the poor, invest millions in new pipes and repair infrastructure. Connections for those who could pay increased, while the situation for the poor, who no longer had access to public water, got worse. Water rates soared by 35 percent and the poor now had their water metered. Seventy percent of the poor of Jakarta still have no running water. In February 2007, the *Jakarta Post* reported that Suez and Thames Water had failed to meet their promised investments and that the rate of new annual hookups under their regimes had been dramatically lower than under the previous public system (down from 11.68 percent a year from 1988–1997 to 5.61 percent a year since then). The company and the government have met stiff resistance and steady documentation of their failures by Jubilee South, a large network of civil society groups working on debt cancellation, and the Indonesian Forum on Globalization, led by tireless activist Nila Ardhianie.

The Philippines

In Manila, water privatization has only exacerbated a system that favors the rich and builds in class disparities. In 1997, with substantive funding from the World Bank and the Asian Development Bank, Manila partnered with several private companies, including Suez, to provide private water services. The new company, Maynilad Water Services, got the concession with a set of supposedly iron-clad promises: lower water fees; uninterrupted water services to existing customers; expansion of services leading to universal coverage by 2006; major reductions of water loss from leaky infrastructure; and compliance with World Health Organization water and effluent standards by 2000. None of these commitments even came close to fulfilment, report critics such as the Water for People Network of the Philippines, which has waged a powerful fight against the companies. Water services deteriorated to seven million of the city's poorer people, and the company started raising rates almost immediately. In October 2003, the west zone of the city suffered a cholera outbreak in which six people died and another six hundred were hospitalized. Subsequent tests by the University of the Philippines showed that Maynilad's water was contaminated with E. coli. Between 1997 and 2007, water rates rose 357 percent.

Australia

Australia's politicians are in denial about the seriousness of their water crisis. The Australian government continues to perceive and sell Australia as a wealthy exporting country with no limit on possibilities for growth and further industrial production. At the very moment that politicians should be coming together in an intensive conservation effort, they are allowing the massive selloff of their water resources. Coca-Cola plants are springing up around the country; private brokers are selling rural water rights; virtual water trade is growing; desalination plants are being built; and the big European companies are running the water systems (badly) in several cities. Fifteen months after Adelaide contracted United Water, a joint venture of Thames Water and Veolia, to run its water supply in 1996, the city was engulfed in a terrible stench the residents called the "big pong." An independent investigation showed it was due to the failure of the company to care for and monitor one of the major sewage lagoons. Between 1993 and 2000, water tariffs in Adelaide increased by 60 percent. In 1998, the residents of Sydney were forced to boil their water when it was contaminated with parasites. The government laid the blame directly at the feet of Suez.

Fierce opposition in Queensland has for years held up plans to recycle sewage water and send it back into the system for drinking and cooking. Citizens Against Drinking Sewage (CADS), organized by plumber Laurie Jones and feisty activist Rosemary Morley, lead the "no" side in a July 2006 referendum in the parched city of Toowoomba, a fight that made international headlines. Queensland premier Peter Beattie has not given up and has plans for other cities. CADS put out a poster in February 2007, when Beattie was photographed drinking NEWater from Singapore with the caption "Why would a Queensland premier be drinking sewage water from Singapore?" When a similar plan to introduce recycled water into Brisbane was announced, CADS circulated four hundred thousand copies of the book *Think Before You Drink: Is Sewage a Source of Drinking Water?* door to door. In Melbourne, Liz McAloon of the Victoria Women's Trust runs an education program called Watermark Australia, a civic engagement project to allow ordinary people to become involved in planning for the future of water.

Others

Similar stories of resistance are heard from other Asia-Pacific countries. Vietnam terminated a wastewater contract with Suez in 1997. In April 2007, the Korean Government Employees' Union and the civil society group Joint Action Against Water Privatization published a report condemning corporate involvement in the water sector in South Korea, to wide public approval. In Malaysia, the Malaysian Coalition Against Privatization, a coalition of 127 human rights, community and environmental groups, went on to lead such a powerful resistance to planned privatization laws that the government gave in, and in January 2005, rescinded its planned water bill and declared the country's water to be a public service. However, their victory was somewhat set back in 2006, when the government passed legislation allowing for thirty-year concessions for the total private management and control of three major heavily populated rivers in the country.

Similar resistance in Sri Lanka held back privatization until the tsunami of December 2004, which devastated the country's water services systems. In order to get the necessary funding and loans to rebuild, said the Asian Development Bank, Sri Lanka would have to accept private sector management of the project. And so, four days after the tsunami struck, the government passed legislation opening up its water sector to privatization. In Nepal, activists passionately opposed their government's 2006 contract with British water company Severn Trent, to run the water supply services in Kathmandu, a contract strongly promoted by the Asian Development Bank. At her April 2007 swearing in, Hisila Yami, the new minister for physical planning and works, criticized her own government's decision to privatize its water supplies, saying that it was akin to selling your mother. The next month, in the face of fierce resistance, Severn Trent announced that it was pulling out of Nepal. Yami immediately promised to keep Kathmandu's water under public control.

Africa

Africa's water desperation is matched by its poverty and the companies have been much slower to take a chance on that continent. However, both privatization and resistance to it have been growing.

South Africa

When apartheid ended in South Africa in 1994, some fourteen million of the country's forty million citizens had no access to water and twenty-one million had no access to sanitation. It was a key promise of the new ANC government under Nelson Mandela to provide water to these (mostly black) communities, a promise that got a start with a pledge to provide each family with six thousand liters of water for free each month. However, under pressure from the World Bank, and in keeping with the new government's commitments to develop South Africa along market-based policies, Suez was brought in to manage Johannesburg's water services and immediately implemented a full-cost system of payment and installed water meters in people's homes. The Mbeki government claims that millions more now have access to water than a decade ago but does not say that many of them – as many as ten million in 2001, according to studies at Witwatersrand University – have been disconnected because of inability to pay under the for-profit system now in place.

A powerful and very vocal opposition to water privatization has sprung up in the cities and townships. The South African Coalition Against Water Privatization is made up of many human rights groups, workers and environmentalists, including the Landless Peasants Movement, the South African Municipal Workers' Union and Jubilee South Africa.

Others

The Namibian Natural Society for Human Rights has been fighting prepaid water meters in Namibia since their installation in 2000, arguing that it puts the price of water out of reach of the majority.

The Bread of Life Development Foundation/WaterWatch in Nigeria has launched a campaign against government support for water privatization, charging that among other irregularities, there has been no environmental assessment of the projects. Nigerian groups are particularly incensed that the government was forced to cut public water access in order to set the stage for private water control. In Gabon, Veolia did at first increase the number of hookups marginally, but investments, which were largely paid for with aid money and the government's own coffers, were nowhere near enough to keep up with the need. When the country experienced its first-ever typhoid outbreak in December 2004, local authorities pointed the finger at the failed privatization experiment. In 2005, the government of Mali renationalized the water system after a poor performance from the French utility SAUR. In February 2007, the government of Guyana canceled a twenty-year contract with British water company Severn Trent after just five years because of broken promises.

A fierce battle has been going on in Ghana for years, with the Ghana National Coalition Against the Privatization of Water and the public sector workers on one side, and the World Bank on the other. The World Bank had laid out conditions for water services funding to the government; after five years of intense struggle, the contract to manage Accra's water system was finally given to Vitens, a Dutch company, and Rand Water of South Africa in November 2005. A 2002 international fact-finding mission noted an extraordinary rise in water rates even before the contracts were signed as the government had already started preparing the people for a regime of full-cost recovery. Today, water services are out of range for the poor. Local civil society groups have had more luck in Tanzania, where the government canceled a contract with British company Biwater in 2005, after only two years. The World Bank put US$143 million into this African "flagship." But the government charged that the company broke many promises, including new pipe installations, investment in water quality and guarantees for more equitable water services. In April 2006, Biwater launched an arbitration dispute at the International Centre for the Settlement of Investment Disputes demanding US$25 million from Tanzania for canceling the project. In a great example of international solidarity, human rights groups from Canada and Switzerland joined Tanzanian civil society groups and filed a joint submission to the court in order to testify against the company.

The United States and Canada

Unknown to many in the global South, similar struggles are going on in the global North. While water privatization has not been forced on Canada and the United States by World Bank structural adjustment policies, the political climate for privatization is ripe as governments embrace market-based solutions for essential services and resource development. As well, cash-strapped municipalities are looking for ways to offload responsibilities and programs. The promise of savings from private investment and water pricing has tempted many municipal politicians to turn over their water systems to private companies, both domestic and foreign. However, in both countries, the public has become very attached to quality water services delivered on a not-for-profit basis at affordable prices and has resisted the selloff of their public water systems to private companies with surprising ferocity.

Canada
Only a very few municipalities in Canada have attempted to privatize their water, and they were met with a strong backlash by a national coalition called Water Watch, founded by the Canadian Union of Public Employees, the Council of Canadians and the Canadian Environmental Law Association, but now broadened out to include students, indigenous groups and faith-based communities as well. In Quebec, the network Eau Secours has been very successful in putting the issue of water on the public agenda. The coalitions successfully stopped planned privatizations in Montreal, Quebec, in 1999, after what one newspaper called "a vast public debate"; Vancouver, British Columbia, in 2001, where more

than a thousand people showed up at a public forum to protest the plan to privatize the city's filtration plant; Toronto, Ontario, in 2002, where the argument that NAFTA would make the decision to privatize water irreversible helped cement the vote to maintain a public system; Halifax, Nova Scotia, in 2003, when Suez, the company the city was negotiating with to clean up the harbor, refused to comply with environmental standards; and Whistler, British Columbia, in 2006, the site of the 2010 Winter Olympics. Critics have been less successful in Hamilton, Ontario, and Moncton and Sackville, New Brunswick, all of which have various levels of water privatization, though opposition remains strong and vigorous.

Canada's larger concern is the threat of large-scale commercial water exports to the thirsty United States. Water is considered a tradable "good" in the North American Free Trade Agreement (NAFTA), which means that if any province allows commercial exports of water to commence, it will be very difficult to turn the tap off. Water is also an "investment," which means that American water corporations (or the U.S. subsidiaries of the big French corporations) could sue the Canadian government for damages if it ever changed the rules and tried to assert control over Canadian water after the companies had set up a commercial presence in Canada. So activist groups in Canada have paid a lot of attention to attempts to export Canadian water for profit, knowing this would trigger the NAFTA process. Water activists have successfully stopped the commercial export of water from the Great Lakes, British Columbia and Newfoundland.

The United States
In the United States, a number of experiments with privatization have been turned back thanks to strong local groups, many of whom have formed a national network called Water Allies, under the leadership of Wenonah Hauter and her team at Food and Water Watch. Hauter is promoting a Clean Water Trust Fund that would finance water infrastructure repairs and points to polls that show that almost 90 percent of Americans would

support such a fund through their taxes. A large activist Listserv called Water Warriors keeps North American and international campaigns in constant contact. As well, Shiney Varghese sends out volumes of materials on the global water struggle from the Minneapolis-based Institute for Agriculture and Trade Policy.

Atlanta, Georgia, signed a US$428 million contract with United Water in 1999, but severed it just four years later citing broken promises, faulty infrastructure and dirty water. New Orleans, Louisiana, dropped its US$1.5 billion contract with Suez and Veolia in 2004 after five years and almost US$6 million worth of study. The companies were balking at new laws giving the voters the right to approve or deny such contracts. Laredo, Texas, terminated its 2002 contract with Suez's United Water in 2005, when the company asked for an additional US$5 million in unexpected expenses. Stockton, California, canceled its contract with Thames and the U.S. wastewater company OMI after a multiyear struggle led by the Concerned Citizens Coalition of Stockton. The citizens of Felton, California, coming together as Felton FLOW (Friends of Locally Owned Water), voted in 2005 to raise their own taxes in order to buy back their water from RWE's water subsidiary, CalAm.

The struggle is still going on in Lexington, Kentucky, where the group Bluegrass for Local Ownership of Water, or Bluegrass FLOW, lost a November 2006 referendum to have their water returned to the public domain. The company American Water, an RWE subsidiary, spent millions on this fight, helping elect pro-privatization candidates in the municipal election of 2004. A company publication explained, "If the primary avenue of attack is legal, the principal line of defense, and the obvious point of counterattack, is political."

Other battle lines revolve around plans to capture, store, move and sell water in water-stressed areas of the country, particularly California. In 2002, angry citizens stopped in its tracks a bid by Alaska water marketer Ric Davidge to ship water from three rivers in Northern California in giant, floating water bags to Southern California. They also put a stop to a plan by the Cadiz Corporation to store, mine and sell water in the Mojave Desert.

The company's stock plunged. Fifty kilometers southwest of Las Vegas, the Sandy Valley Water Warriors celebrated a state Supreme Court victory in November 2006 over the Vidler Water Company's bid to take fourteen hundred acre-feet of water from the Sandy Valley Basin and pump it over a mountain to sell to desert developers. The town, with fewer than two thousand residents, raised us$60,000 for this case. A new group, the Progressive Leadership Alliance of Nevada (PLAN), has come together to fight the proposed pipeline that would ship water from rural Nevada to Las Vegas. This would be the largest transfer of water in the United States, if successful. But given the passion of those fighting to prevent this major mistake, it would be unwise to assume the die is cast. Another California group, the California Water Impact Network (C-WIN) led by Carolee Krieger, is using the courts to challenge privatization and deregulation of the state's water supplies as well as the reliance of developers on "paper water" – water that exists only in contracts and not in the ground.

Europe

Thankfully, a strong movement to protect public water rights has emerged in Europe as well, the home of the big water transnationals. The two overarching goals are to force the European Union to stop funding privatized water services in the developing world and to prevent water from being part of a single EU market, which would harmonize and institutionalize market-based competitive water services in Europe. Individual struggles are going on in most countries.

In Ireland, for example, opposition is growing to a plan to install pre-paid water meters. The citizens of Sicily, Italy, are fighting to wrest control of their water from the Mafia, who have benefited from its privatization. The citizens of Herten, Germany, bought all the shares in their water company when it was put up for sale. In April 2004, environmentalists and water activists in Spain celebrated when the government abandoned plans to build a gigantic pipeline and pump water from the Ebro River in the north to the thirsty cities of the south.

Italian academic and visionary Riccardo Petrella initiated a water project coming out of the influential Lisbon Group, called the International Committee for a Global Water Contract. In March 2007, he convened the World Water Assembly for Elected Representatives and Citizens (AMECE in French) with more than five hundred activists, academics, journalists and politicians who committed themselves to a more rigorous and specific global program to ensure access to water for all. Petrella works closely with the elegant Danielle Mitterrand, widow of former French president François Mitterrand and head of France Libertés, a foundation dedicated to the equitable distribution of water and working to return French cities and towns to public water systems. Mitterrand has paired communities in France with communities in Bolivia to support Bolivia's struggle to build public water services in the wake of the failed privatization schemes. In 2006, Germany's Bread for the World and the Heinrich Böll Foundation helped set up the World Council of Churches Ecumenical Water Network to promote the "preservation, responsible management and equitable distribution of water for all, based on the understanding that water is a gift of God and a fundamental human right."

Also key to the movement is the research done by David Boyes and his team at Public Services International and David Hall at the Public Services International Research Unit. They provide critical data and analysis not found anywhere else. Olivier Hoedeman and others at the Amsterdam-based Corporate Europe Observatory (CEO) also publish excellent research, expose corporate-political ties and keep up constant pressure on European politicians to be transparent. On the 2007 International Water Day, CEO and sixty other groups published an open letter in the *European Voice* critical of the European Commission's promotion of water privatization in the Third World. CEO and PSI work closely with the British-based World Development Movement, Netherlands-based Transnational Institute, Friends of the Earth International and the international network of NGOs and grassroots groups opposed to the World Trade Organization called Our World Is Not for Sale. Together, they were responsible for forcing the WTO to drop drinking water from the GATS negotiations on services. Working with activist groups in Nepal, the World Development Movement led the successful campaign from England to get the British water company Severn Trent to withdraw its water privatization bid in that country. The Norwegian Association of International Water Studies (FIVAS) was instrumental in getting the Norwegian government to pull its support for the Public-Private Infrastructure Advisory Facility (PPIAF) that continues to fund private water services, and to tell the World Bank it will no longer fund any water program that is not public.

A Global Water Justice Movement Is Born

While all of these struggles have been waged in these various countries, it has been crucial to develop national, regional and international networks to link strategies, share research and provide financial and resource solidarity where possible. Today, a coordinated and highly effective international water justice movement is fighting both the power of the private water companies and the abandonment by their governments of the responsibility to care for their national water resources and provide clean water to their people. Using interactive websites and Listservs such as Water Warriors, this network is able to get hundreds of groups to sign on to a petition or a demand with twenty-four hours notice. Much of the early work of these networks took place at international meetings – often of the very institutions we oppose.

Second World Water Forum – The Hague, March 2000
Galvanized by these first local struggles, civil society groups from a dozen countries headed to the Second World Water Forum in The Hague in March 2000, where we met as the Blue Planet Project. Although we were not part of the official agenda, we

gathered in any unoccupied rooms we could find to create our own Vision Statement to counter the Vision Statement of the World Commission on Water for the 21st Century. In it, we expressed "serious concerns" about the process and content of the World Water Council's Framework for Action, which we accused of being "dominated by technocratic and top-down thinking, resulting in documents which emphasise a corporate vision of privatisation, large-scale investments, and biotechnology as the key answers." The process gives "insufficient emphasis and recognition of the rights, knowledge and experience of local people and communities," we said, "and the need to manage water in ways that protect natural ecosystems, the source of all water." We called for water to be considered a universal right and to be removed as a tradable good from all trade agreements.

We took this vision to the media and into the great meeting halls where we told the local stories of struggle to the assembled participants from the mics set up around the room. On one occasion, when a World Bank official who was chairing a plenary in front of an audience of thousands refused to recognize the dissenters at the mics, I lined up a dozen protesters at a floor mic and announced that I was now co-chairing the session and told the unhappy (but temporarily powerless) moderator that we would be hearing from these twelve people before he went back to his speakers' list. The press loved this, of course, as did hundreds of conference participants in the audience who shared our views. While we did not sway the official outcome of the summit, we put the powers of the World Water Council on notice that we had arrived as a movement and would not be going away.

World Bank Protest – Washington, April 2000

A mere month later, our fledgling movement took our message to Washington for the annual meeting of the World Bank, where we marched with the many thousands in the streets against World Bank policies and raised the issue of water privatization for the first time. The International Forum on Globalization, a San Francisco–based policy and research institute critical of globalization, held a large teach-in at the Foundry Methodist Church. Oscar Olivera left Bolivia (for the first time) to attend the teach-in and march with the protesters from all over the world. He was picked up at the airport and whisked at high speed to the packed evening event where a grateful audience greeted him, many of whom were in tears, with a prolonged standing ovation for his courage and leadership and to show international support for the Bolivian water revolution.

Water for People and Nature – Vancouver, July 2001
In July 2001, the Council of Canadians, a national Canadian public advocacy organization, and the Blue Planet Project housed at the council, hosted Water for People and Nature, the first global civil society summit of water activists. More than eight hundred activists, academics, environmentalists, human rights experts, indigenous peoples and public sector workers from forty countries came together in Vancouver, British Columbia, to launch a coordinated international network of grassroots activists and form national and international organizations to fight for the preservation of the world's water and the right to water for all people. The plenary voted unanimously to oppose the "fake logic" of the marketplace for water distribution and called for a UN convention to protect water as a human right and as part of the global commons. It was agreed that we would build a working relationship based on the understanding that groups from the global South had much to teach groups from the global North and that the movement would be founded on the principles of equality and solidarity, not charity and "development."

The most moving moment of this seminal event was a moment's silence for Kimy Pernia Domico, an activist from Colombia who was to have spoken at the conference but was "disappeared" just weeks before. Kimy was a leader in the struggle against the Urra hydroelectric dam, on his people's traditional lands, that was destroying their livelihoods. He knew his opposition to this project had put his life in danger. As he said in his 1999 testimony to the Canadian Parliament, "Saying these things to you today puts my life in danger. The paramilitary gunmen have set fire to our boats to prevent us from going to meetings. . . . Anyone who dares to speak out about Urra is accused of being involved with the guerrillas and they have declared both our communities and leaders to be a military target." As we sat praying for Kimy, we were keenly aware of the danger of this work for many of our colleagues and the courage it takes for someone like Kimy to carry on in the face of such threats. (Tragically, in January 2006, Salvatore Mancuso, a senior commander with Colombia's right-wing militias, admitted to murdering Kimy Pernia Domico, along with hundreds of others during his fifteen years with the death squads.)

World Summit on Sustainable Development –
Johannesburg, August 2002

The next major opportunity to gather in force was the World Summit on Sustainable Development in Johannesburg, August 2002. Once again the International Forum on Globalization mobilized a large teach-in, this one at Witwatersrand University. This event brought together thousands of African activists, academics and environmentalists, as well as leaders of the movement, including Virginia Setshedi, Patrick Bond and Trevor Ngwane from South Africa; Tewolde Egziabher from Ethiopia; and Patrick Apoya and Rudolph Amenga-Etego from Ghana. Here, the heart-wrenching stories of corporate water theft and the devastation of whole communities galvanized the movement. Participants were highly critical of the South African government of Thabo Mbeki, who not only openly supported the corporate-dominated WSSD summit but also had recently introduced private water delivery in Johannesburg and several other communities in South Africa, which had led to water cutoffs to many thousands.

We rented an old school bus and took teach-in participants to Orange Farm, another township of terrible poverty, guests of the Orange Farm Water Crisis Committee and its fearless leader, Richard (Bricks) Mokolo. As far as one could see were burning tires and garbage, rats in the streets, pit latrines and no running water in the small corrugated-iron hovels the people call home.

There, we were shown the brand-new water pipes, one per block, that had recently been installed by Suez and the state-of-the-art water meters attached between the pipes and the taps. Every drop of water must be paid for in this and other townships where the majority are too poor to pay, giving new meaning to the saying, "Water, water everywhere and not a drop to drink." As a result, residents – mostly women – must walk many kilometers to fetch water from rivers and streams with cholera-warning signs lining their banks. As we inspected the new pipes and heard the stories of children dying of dirty water from local residents, a huge, double-decker BMW bus pulled up and out stepped dozens of beautifully dressed senior officials from the European delegation of the WSSD as well as VIPs from Suez, out to impress the politicians and bureaucrats with their new pipes. As soon as it became clear who these new visitors were, angry residents chased them back onto their bus and chanted them away as the driver careened out of sight.

So incensed was Mbeki over criticism of the WSSD (he publicly railed against those opposed to the summit and the presence of the water corporations as "anti-poor"), his government threatened to pull the permit for the up-till-then legal United Social Movements march to be held on the final day of the summit. To protest this threat, earlier in the week we led a peaceful candlelight march of seven hundred men, women and children out of the teach-in at the university and out of the university grounds. We were immediately met by a phalanx of riot police who threw stun grenades into the crowd, creating panic and wounding several marchers. Because of the presence of international speakers, the international media caught all this on film, and terrible images of police brutality against peaceful protesters went around the world. Not only did this lead to an outpouring of criticism of these heavy-handed tactics in the South African media, it opened the door to our views in the mainstream newspapers and helped discredit the WSSD and its corporate orientation internationally.

As well, the Mbeki government was forced to permit the big march on the final day, a gathering of more than twenty thousand that streamed out of the Alexandra slum and wove its way down an eight-lane highway to the site of the summit, the ultra-wealthy Sandton Business Park. There, in front of the world's media, the people called for access to water, life and dignity and an end to the economic apartheid created by water privatization.

Third World Water Forum – Kyoto, March 2003

By the Third World Water Forum held in Kyoto in March 2003, the international and very public criticisms of water privatization and of the World Water Council had hit some important nerves, and civil society critics were asked to participate inside the summit for the first time. Our strategy, planned with civil society groups in Japan, was to accept all formal venues to present our alternative vision, to bring our colleagues from the global South to tell their stories to the participants and media and to work with Japanese civil society to build a strong support for public water systems, of which Japan has one of the best in the world. We met with proud public sector water managers from different prefectures who showed us their technical expertise in caring for Japan's public water services. Tony Tujan of the IBON Foundation in the Philippines made the obvious observation that, with Japan's public sector expertise so close, it would have saved everyone a lot of money and grief to bring Japanese water specialists to Manila to transfer their knowledge and experience.

I was asked to co-chair, with the World Water Council, a major theme session on public-private partnerships and, at the end of an intensive two-day debate, I submitted our independent report totally opposed to these partnerships. This meant that, on the most contentious issue of the forum, the position of civil society had to be officially placed into the record of the final transcript. We arrived with our own vision statement signed by three hundred organizations around the world and, under the rallying cry "Water Is Life," swarmed the mics in every session with stories from the grassroots. At a forum with the heads of the big water companies all sitting on the stage, a worker from Cancun held up two plastic bottles both filled with water supplied by Suez.

One was from the five-star hotel where he works and was crystal clean; the other from the barrio in which he lives and from where he commutes to work. The water in that bottle was brown and putrid-smelling. He challenged Jean-Louis Chaussade, CEO of Suez Environment, to drink "his" water from both bottles; Chaussade declined.

When former IMF director Michel Camdessus launched his controversial report on water financing, *Financing Water for All*, we held up hundreds of "lie meters," brightly colored half-moon signs with an arrow to indicate the strength of the lie and little bells attached that we rang quietly or more strongly depending on the statement. At one point, Camdessus, clearly agitated, looked up and said, "I hear your tinkling! It won't stop me."

Red VIDA – El Salvador, August 2003

An important milestone in our movement was the founding of a network of the grassroots groups of the Americas called Red VIDA, which in English translates as Inter-American Vigilance for the Defence and the Right to Water. The new network came out of a regional seminar held in El Salvador in August 2003, where it was agreed that a more formal organization was needed to coordinate the activities of all the groups fighting for their water rights and better organize the campaign to stop the privatization of the hemisphere's water resources. As in other gatherings, we took the time to articulate our shared principles and values, calling for social, sustainable and universal water services, and the understanding that water is "a public good and an inalienable human right to be protected and promoted by all who inhabit the planet."

The first assembly of Red VIDA took place in Porto Alegre, Brazil, on January 25–27, 2005, just prior to the World Social Forum. Thirty organizations from fourteen countries came together to formulate alternatives to private water control and launch an international campaign against Suez. Many of the groups traveled to Cochabamba, Bolivia, that August to show public support for the struggle there as well as the campaign in La Paz and El Alto against Suez, and then came together to support the plebiscite in Uruguay that led to a successful referendum to have water declared a human right during the national election of October 2004. On March 24–26, 2007, the second Red VIDA assembly took place in Lima, Peru, hosted by FENTAP, the Peruvian water workers' union, where forty organizations from fourteen countries gathered and advanced their common work plan to promote public alternatives to private water management.

Red VIDA members have been key components of the global campaign against Suez. VIDA members from Bolivia, Argentina, Uruguay and Chile joined water activists from the Philippines outside the May 13, 2005, Suez shareholders meeting in Paris to protest the company's practices. While some surrounded the building with colorful banners, members of the Boston Common Asset Management, a U.S. socially responsible investment firm that holds shares of Suez and has been critical of its practices, read out a statement denouncing the company inside the meeting. Similar peaceful protests took place outside Suez headquarters in Buenos Aries, Quito, La Paz, Montevideo, Manila, Rome and other cities.

People's World Water Forum – Delhi, India, January 2004

Another venue that has come to be very important to the global water justice movement is that of the World Social Forums, an annual global gathering of activists, environmentalists, academics and progressive politicians who meet as a counterpoint to the annual World Economic Forum in Davos, Switzerland, attended by global political and business elites, and to promote alternatives to economic globalization, privatization and corporate domination of the economy. The First World Social Forum took place January 2001 in the city of Porto Alegre, Brazil, which also hosted the events for the next two years and again in 2005 (this one attended by 150,000 people from around the world). At each of these gatherings, our water movement has held workshops, "speak-outs" on private water resistance, and strategy meetings to promote our work and strengthen our networks. In 2004, the World Social Forum was held in Mumbai, India, where we also met extensively as a movement.

But we also held a separate assembly, the People's World Water Forum in Delhi, January 12–14, just days before the big gathering in Mumbai, where water activists from sixty-five countries converged on the India International Centre in Delhi, organized by world-renowned food and water scientist Vandana Shiva and her Research Foundation for Science, Technology and Ecology. This gathering was particularly important to support the movements of the Asia-Pacific and to tell people of the region that these struggles are going on all over the world. We were deeply moved to meet and support Rajendra Singh, known as "the water man," who has led a movement to harvest rainwater in Rajasthan. His organization, Tarun Bharat Sangh, has worked with a thousand villages to pull themselves out of a drought-related crisis. The proponents of private water hate him. In 2002, Singh was ferociously beaten by thugs generally believed to be connected to local authorities.

Out of the Delhi summit came a renewed conviction to work against the World Bank; to get water removed from the General Agreement on Trade in Services; to campaign collectively against both Coca-Cola and Suez; to fight for a UN convention on the right to water; and to create a new network based on the need to provide and support alternatives to private water services. The Reclaiming Public Water network is an international civil society network sharing information, strategy and resources in order to promote public-public partnerships and water operator partnerships, whereby public water utility expertise is transferred where needed and public systems are supported by public funds from the global North such as public pension funds.

Fourth World Water Forum – Mexico City, March 2006
Mexico City was the site of the Fourth World Water Forum in March 2006. Rather than try to influence the official forum, the water justice movement decided to stage its own events, beginning with a thirty-five-thousand-strong march held the first day of the World Water Forum. One thousand activists and academics also attended an alternative peoples' forum, the International

Forum on the Defense of Water, organized by the Mexican human rights coalition COMDA. A highlight of this forum was a presentation by Bolivia's new water minister, Abel Mamani, who shared his "human vision" of water with us and promised to support the right to water both in his own country, within Latin America and at the United Nations. Our movement held a huge rally and concert in the enormous Plaza Zócalo in Constitution Square, where I gave a speech to twenty thousand young people on the right to water. Our message drowned out the official forum message in the media and, when water activists from around the world left to go home, we knew we were leaving behind a strong and energized movement in Mexico.

Seventh World Social Forum – Nairobi, January 2007

On January 27, 2007, 250 grassroots activists from more than forty African countries came together in a packed room at the gigantic Moi Stadium in Nairobi, Kenya, to form the African Water Network, the first pan-African network to coordinate efforts to protect local water sources and stop the corporate theft of their water supplies. For many of us who have been involved with the struggle for a long time, this was a very moving moment. Co-chairs Al-Hassan Adam of the Ghana Coalition Against the Privatization of Water and Virginia Setshedi from the South African Coalition Against Water Privatization warned their governments and the World Bank that the abuse had to stop. "Today we celebrate the birth of this network," said Setshedi, "tomorrow, access to clean water for all." Added Adam, "The launch of this network should put the water privateers, governments and international financial institutions on notice that Africans will resist privatization."

The highlight of the World Social Forum for me was a fact-finding trip to the haunting Lake Naivasha, home to one of the last major wild hippopotamus herds in East Africa, site of the Robert Redford/Meryl Streep movie *Out of Africa*, and now on the brink of extinction in order to supply roses for Europe. The shallow lake, a UN Ramsar (wetland protection) site, lies surrounded by volcanoes on the floor of Kenya's Great Rift Valley, and is a paradise of biodiversity, flush with giraffes, zebras, water buffalo, lions, wildebeests and at least 495 species of birds. Until 1904, when the government signed an agreement opening it up to European settlers, Lake Naivasha and the land around it were protected to provide grazing and hunting for the Maasai people. Soon, European settlers had bought up all the best land, building plantations on the shoreline and then surrounded these homes with a ring of flower farms. The population started to grow and then exploded, from 7,000 in 1985 to more than 300,000 today, to service the flower industry. The vast majority of the workers – all black and mostly female – and their families live across the road from the farms in slums with no running water and pit latrines that leach into the lake.

Kenya is the largest producer of cut flowers in Africa and the leading supplier to Europe. Britons alone spend US$3 billion a year on cut flowers, and Kenya supplies one-quarter of that market. About thirty major growers, almost two-thirds of them foreign owned, surround the lake with huge industrial farms, closed to the public with iron gates and armed guards. (Sexual abuse is rampant, many workers make only US$1 a day, and many are ill from the heavy use of pesticides and herbicides. We drove around the back of one farm to the workers' entrance where a sign warned employees that their lives would be in danger if they removed company property.) Roses are 90 percent water, and Europe is using this and other African lakes to protect their own water sources from exploitation. The results are catastrophic: the lake is half the size it was fifteen years ago and the hippos are literally dying in the parched sun. If nothing is changed, the lake will be a "putrid muddy puddle" in ten years, say scientists. As a result of the visit, a new network, Friends of Lake Naivasha has formed, to save this enchanted place. But Lake Naivasha is only one of dozens of African lakes being drained for profit – the latest legacy of a colonial relationship not yet past.

Bottled Water Warriors

An energetic international movement is coming together to challenge the bottled water industry as well. While this movement understands that people in many parts of the world do not have access to clean public water, and are therefore forced to turn to bottled water, it is the long-term hope that the day will come when the world's surface water sources are clean and accessible and bottled water will become a thing of the past. It also challenges the use of bottled water in many countries where the public water is not only clean but more regulated and likely safer than water in bottles. And perhaps most importantly, this movement challenges the growing corporate control of not only water but also water policy by the big bottled water companies.

A case in point: In 1998, Nestlé chose Pakistan, only one-quarter of whose citizens have access to clean water, as the country to roadmap its global water strategy in the bottled water market and convinced the government of Pakistan that bottled water was the answer to the country's water crisis. It introduced its new product, Pure Life, as the only real source of clean water that provides minerals, helps prevent obesity and reduces the risk of health-related problems. The Pakistani government granted Nestlé access to several large aquifers and the company's profits soared as the consumption of bottled water jumped 140 percent in two years. However, the relationship soon soured as water tables dropped, and it became obvious the company was draining future water sources for profit and violating its own human rights commitment made as part of the United Nation's Global Water Compact. As well, reported Nils Rosemann in a 2005 report for the Swiss Coalition of Development Organizations, it soon became clear that the price of Pure Life was out of the reach of the majority. Reacting to a strong public outcry and an organized anti-Nestlé citizens campaign, in February 2005, the Pakistani government served notice to Nestlé management that it was selling its products without authorization. The two sides have been tied up in court since then.

Similar resistance has risen all over the world. Franklin Frederick of Brazil's Citizens for Water movement traveled all the way to Nestlé's headquarters in Vevey, Switzerland, in June 2005 to protest the damage it is doing in his hometown of São Lourenço, where overpumping has destroyed the taste of the ancient and famous mineral waters of the region. Charges of foul odors have plagued Coca-Cola in the village of Barangay Mansilingan, near the port city of Bacolod in the Philippines, where five hundred families have charged the company with dumping harmful contaminates into their water supply. In Indonesia, WALHI, a coalition of the Indonesian Forum for the Environment and Friends of the Earth Indonesia, has led the fight against large concessions the government has given to Danone and Coca-Cola to take massive amounts of groundwater in Central Java, where the companies are destroying the livelihoods of thousands of farmers. Opposition is growing in the war-torn state of Chiapas, Mexico, where Coca-Cola has been granted favorable zoning laws to extract enough water to supply five villages while local residents go without water. Some licenses are granted to 2050.

On January 20, 2005, thousands of people all over India surrounded its eighty-seven Coca-Cola and PepsiCo plants and told the companies to "Quit India" as they were violating the constitutional guarantee of the right to life. All over India, these companies have met intense resistance to their exponential water takings (most of which they get for free or a nominal fee), the most passionate and organized from some of the poorest people on the planet. In June 2006, community leaders in Mehdiganj in north India waged a twelve-day hunger strike in front of the Coca-Cola plant, charging that high levels of cadmium and lead were being discharged from it. More than forty people, including the famous anti-dam leader Medha Patkar, were arrested in Delhi on International Water Day, March 22, 2007, for a peaceful protest over the water shortages being created all over India by these bottling companies.

"The world needs to know that the Coca-Cola company has an extremely unsustainable relationship with water, its primary raw material," said Amit Srivastava of the India Resource Center and a leader of the anti-Coke movement. "Drinking Coca-Cola contributes directly to the loss of lives, livelihoods, and communities in India."

The struggles take a terrible toll. M. P. Veerendrakumar, the leader of the Indian Newspaper Society, describes the years of court battles the people of the village of Plachimada have waged against Coca-Cola. Even though the courts closed the local factory down, "the reality is the cola companies are still continuing their activities, unabated and unchecked. The feeble voice of the people is still listened to by only few. Little is done to end this exploitation, which causes disastrous consequences. The woes of these forsaken people cast a gloom over this accursed land. Their miseries enmeshed in a tangle of escalating litigation and technicalities, the voiceless dalits and adivasis [once called 'untouchables'] wage a last ditch struggle against an inhuman enemy, as the sands slip away from underneath their wobbling feet."

Similar contests are growing in the global North as well. Nestlé has seventy-five spring sites under seven brand names in the United States: Poland Spring, Ice Mountain, Deer Park, Zephyrhills, Arrowhead, Ozarka and Calistoga. A fierce showdown has erupted in the northeast corner of California between the residents of McCloud and Nestlé, which has been given the go-ahead to bottle and sell water from the slopes of Mount Shasta. As well, the company has been given unlimited rights to groundwater in the area and control of the major dam on the McCloud River. Across the country, in Michigan, Sweetwater Alliance and Michigan Citizens for Water Conservation have sued Nestlé for trying to capture and export Great Lakes water through a pipeline it installed on a private ranch it bought next to the lake. Maine has become a major battleground, pitting environmentalists, farmers and activists against Nestlé's Poland Springs in Fryeburg, which got access to a nearby aquifer. Save Our Groundwater (SOG), led by quiet-spoken but deeply determined community activist Denise Hart, has been fighting a huge water taking permit in Nottingham, New Hampshire, which gives a company called U.S.A. Springs the right to take more than 1.6 million liters of water a day out of the ground, threatening farms and businesses in the community. While not successful in stopping the project, the group has forced the state government to bring in protections for groundwater. Next door, in Vermont, the Vermont Natural Resources Council, alarmed at reports of indiscriminate water extractions across the state, has launched a bipartisan campaign toward a similar law.

Groups have made campaigns against these companies both national and international. In North America, the Polaris Institute, Corporate Accountability International (CAI) and Alliance for Democracy have taken the lead in both research on and campaigns against bottled water. CAI holds a Tap Water Challenge on university campuses and in church basements and challenges blindfolded participants to differentiate between tap and bottled water. Most cannot. The Campaign to Stop Killer Coke is another group with an even stronger message: Coca-Cola has violated human rights at its unions in such places as Guatemala, Nicaragua and Colombia, and community rights in India. These and other groups attend the annual Coca-Cola shareholders meetings where they publicly challenge the company and garner a lot of media. At the 2004 meeting in Wilmington, Delaware, activist Ray Rogers was wrestled to the floor and carried out by security guards when he refused to stop speaking. They also tour university and college campuses calling for boycotts. More than one hundred colleges and universities in the United States alone have anti-Coke programs in place and at least twenty have banned Coke outright. So controversial is the company that it was forced to pull out as a core sponsor of the 2005 Live 8 concerts in the wake of global criticism. And in July 2006, KLD, the leading marketer of "corporate social responsibility" in the United States, removed Coca-Cola from its index of socially responsible companies, citing persistent problems with labor practices in overseas plants, marketing practices toward children and water abuse in countries such as India.

The Bottling Companies Fight Back

This organized resistance to the power of the big bottling companies and the environmental and social destruction they have created has forced them to launch public relations campaigns to counter the damage to their corporate images. As part of its Commitment to Water, Nestlé sponsors the Montana-based Water Education for Teachers (WET), which publishes a curriculum for schools on water-related issues such as how watersheds work and the connection between clean water and health, and has trained more than 180,000 teachers in twenty-one countries to run their program. Project WET ran the Children's World Water Forum at the 2006 Fourth World Water Forum in Mexico City, involving children in this controversial gathering, and throwing its support behind the World Bank and the corporations of the World Water Council.

Starbucks funds projects to bring clean drinking water to poor communities through its Ethos Water brand. The company now sponsors a Walk for Water every year on International Water Day to "help children get clean water," and gives money to water projects in India and Kenya. "By purchasing Ethos™ water, customers have been part of this opportunity to make a difference in the lives of children and their communities around the world," said Jim Donald, president and CEO, on International Water Day 2007, on the Starbucks website. (Activists were furious at the cynical attempt by a corporation to hijack this important day in order to promote its logo and brand. One put up an alternative ad on the Ethos website that read, "Did you know that we lure our customers into buying our water by selling them on the idea they are helping the world's children? It's a fact: Ethos Water sells for $1.80 per bottle in your local Starbucks store and only 5 cents goes toward the goal. . . . That's right. We're going to make $360 million selling water on the promise that we are helping the world's children.")

Hoping to restore some of the goodwill that made its flagship brand a global icon, reports the *Wall Street Journal*, Coca-Cola has gone on a clean water kick in the developing world. The company has some seventy clean water projects in forty countries, a service it hopes will help it counter the global anti-Coke campaign as well as secure new customers. The company is also hoping to cure a public-relations headache, notes the *Journal*, caused by its own thirst for water. Its four hundred beverage brands use more than 280 billion liters of water per year. Company chairman E. Neville Isdell says "water stewardship" is now priority number one. Coca-Cola has teamed up with the Blood: Water Mission project of the rock band Jars of Clay, which sends a percentage of revenues from its *Good Monsters* CD to provide clean water in Africa. Coca-Cola has also teamed up with Cargill, Dow Chemical and Proctor & Gamble as well as UNICEF and CARE to set up the Global Water Challenge (GWC) intended "to deliver clean water and sanitation and hygiene education" in the developing world. At the June 2007 annual meeting of the WWF in Beijing, Coca-Cola pledged US$20 million for water conservation.

While some of these individual efforts may bring some water to some families and communities, it is very important to see them for what they are – an attempt by the companies to blunt criticism of their conduct with "feel good" charity while making money. "We want to be seen as a friend and supporter of communities where we operate, and this is a great way to have positive relationships with the local communities," Coca-Cola's director of global water partnerships, Dan Vermeer, told *Pink* magazine, who noted that the company could use a good news story or two. "As we do that on a broader basis, there's a story that can have a positive impact on the way people think of and value the company."

But the major problem with this green-washing PR is that it masks the real issue of water abuse and the role that corporations such as Coca-Cola and Nestlé (and Suez and Veolia) play in this abuse. It also puts a human face on a deeply flawed and corporate controlled system of water distribution that, with the few charitable exceptions these companies make to their own rules through these projects, supplies water as a private commodity to those who can afford to buy it and denies it to those who cannot. The few they reach through these charitable acts pale in comparison with the millions they deliberately leave behind in their quest to control the world's water. In their world, water is not a fundamental human right for every person on Earth, but a market product increasingly controlled by the private sector for its own personal profit.

Perhaps if there were unlimited supplies of clean water, this issue would be less crucial. But with the dire warnings now coming from every part of the globe about the coming water shortage and the resultant wars, the unaccountable and undemocratic private control of water is no longer acceptable.

⟞⟍∿⟋⟝

From thousands of local struggles for the basic right to water, galvanized through these international meetings, a highly organized and mature international water justice movement has been forged and is shaping the future of the world's water. This movement has already had a profound effect on global water politics, forcing global institutions such as the World Bank and the United Nations to admit the failure of their model, and it has helped formulate water policy inside dozens of countries. The movement has forced open a debate over the control of water and challenged the "Lords of Water" who had set themselves up as the arbiters of this dwindling resource. The growth of a democratic global water justice movement is a critical and positive development that will bring needed accountability, transparency and public oversight to the water crisis as conflicts over water loom on the horizon.

Chapter 5

The Future of Water

To the thirsty I will give water without price from the fountain.

– Revelation 21:6

The three water crises – dwindling freshwater supplies, inequitable access to water and the corporate control of water – pose the greatest threat of our time to the planet and to our survival. Together with impending climate change from fossil fuel emissions, the water crises impose some life-or-death decisions on us all. Unless we collectively change our behavior, we are heading toward a world of deepening conflict and potential wars over the dwindling supplies of freshwater – between nations, between rich and poor, between the public and the private interest, between rural and urban populations, and between the competing needs of the natural world and industrialized humans.

Water Is Becoming a Growing Source of Conflict

Inside Countries

Water conflicts already range from the intensely personal to the geopolitical. Australia's water crisis is spawning personal animosity and even fights. In Sydney, "water rage" has led some people to become vigilantes in order to stamp out neighbors' illicit irrigation. In the drought-ravaged rural communities, water thefts are on the rise and the fire brigade in Minyip – the town made famous by the television series *The Flying Doctors* – has been forced to padlock its bore tap. The water crisis in the American Midwest has communities and industries along the Colorado

River fighting with one another for its dwindling water supplies. Some US$2.5 billion in water projects are planned or underway in four states, reported the *New York Times*, along the "1,400-mile-long silver thread of snowmelt and a lifeline for more than 20 million people in seven states." Tempers are flaring among competing water users, old rivalries are hardening and some states are waging legal fights, reports the *Times*. Montana has filed a suit against Wyoming for taking more than its share from tributaries of the river, and population growth in Nevada and Utah have those two states at loggerheads over several proposed pipelines.

Tempers are rising in China as rain is "stolen" by cloud seeding. In Henan Province, the *Chinese Daily* reports that five arid cities are in angry competition with one another to capture the rain before clouds move on. In Klaten in Indonesia's Central Java (where Danone is draining the aquifers), water is so scarce that every time the farmers go into their fields to water their crops, they come armed with axes, saws and hammers to fight one another for the dwindling supplies. In the huge slum of Kibera in Nairobi, Kenya, a million people must share only six hundred padlocked pit latrines (which they have to pay to use), so many are forced to defecate into plastic bags, referred to as "flying toilets." Reports of family violence are on the rise in this slum and elsewhere, as women, the ones responsible for finding and using water, come home empty-handed to face angry husbands and fathers.

Women are at the forefront of an intense struggle that sets the indigenous Mazahuas against Mexican authorities who have confiscated their water sources for use in Mexico City. A quarter of Mexico's water has its source on Indian lands, yet many indigenous communities have no access to water. Sixteen thousand cubic liters a second rush by the lands and homes of the Mazahuas, yet eight villages have no water lines whatsoever. A group of women set up the Zapatista Army of Mazahua Women in Defence of Water and on December 11, 2006, illegally shut off the valves of one of the system's plants. In response, the National Water

Commission sent five hundred police to occupy their villages. Beatriz Flores, one of the protesters, told Inter Press Service News Agency that she is prepared to go to jail as the struggle heats up.

In Latin America, anger is growing at the practice of wealthy North Americans and Europeans buying up vast tracts of land, thereby gaining private ownership of the waters on the property. As the *New Internationalist* reports, in the Patagonia region of Argentina and Chile, CNN's Ted Turner owns fifty-five thousand hectares. Benetton magnate Luciano Benetton has bought nine hundred thousand hectares – equivalent to half the area of Wales – and has been challenged by several local Mapuche families who were evicted in 2002 by police after they occupied land on the Benetton estate they say is their ancestral home. A lawsuit is still making its way through the courts. Journalist Tomas Bril Mascarenhas says that fences are marching across the Patagonian wilderness, displacing indigenous peoples and turning private pure water into private property.

Water activists in South Africa routinely teach local communities how to dismantle water meters, an illegal act. In July 2006, Sri Lanka's air force carried out strikes against Tamil Tiger rebels who cut off water diversion to an irrigation channel that interrupted their water supplies. On January 7, 2007, employees of the Nepal Water Supply Corporation, the local public water supplier in Kathmandu, cut water supplies to the royal palace, the prime minister's residence and the city's convention center in a protest against a bill allowing the transfer of the city's water system to the British water company, Severn Trent. (Company employees did not cut supply to the city's residences.) As a result, Severn Trent withdrew its bid. Malaysia passed a new water law in May 2006 providing for capital punishment for those who contaminate water in a way that endangers lives or causes death. More than half the rivers in the country are polluted, reports the *Asia Times*, and the soaring costs of maintaining these systems has caused alarm in government circles. The new law has widespread support in Malaysia. Water is also at the heart of the horrible conflict in Darfur, says Human Rights Watch and others. Desertification and increasingly regular drought cycles pitted nomads against farmers, which led to attacks on government outposts by farmers in the south. Instead of dealing with the water crisis, the government sent in the Janjaweed. "I don't think you can separate climate change from population growth, rising consumption patterns and globalization," writes Michael Klare in his 2002 book, *Resource Wars*. "It's really one phenomenon."

None of this comes as a surprise to scientist Marc Levy of the Center for International Earth Science Information Network at Columbia University's Earth Institute in New York, who has studied the connection between drought and conflict for years. At an April 14, 2007, press briefing in New York, Levy described how he and colleagues used decades of detailed precipitation records, which they overlaid with geospatial conflict information, to come up with a model to track the clear link between water shortages and violence within nation-states. Levy said many domestic water conflict-related deaths have already occurred in the last decade, and singled out Nepal, Bangladesh, Cote d'Ivoire, Sudan, Haiti, Afghanistan and parts of India as ripe for more.

Between Countries

Around the world, more that 215 major rivers and 300 groundwater basins and aquifers are shared by two or more countries, creating tensions over ownership and use of the precious waters they contain. Growing shortages and unequal distribution of water are causing disagreements, sometimes violent, and becoming a security risk in many regions. Britain's former defense secretary, John Reid, warns of coming "water wars." In a public statement on the eve of a 2006 summit on climate change, Reid predicted that violence and political conflict would become more likely as watersheds turn to deserts, glaciers melt and water supplies are poisoned. He went so far as to say that the global water crisis was becoming a global security issue and that Britain's armed forces should be prepared to tackle conflicts, including warfare, over dwindling water sources. "Such changes make the emergence of violent conflict more, rather than less, likely," former British prime minister Tony Blair told *The Independent*. "The blunt truth is that the lack of water and agricultural land is a significant contributory factor to the tragic conflict we see unfolding in Darfur. We should see this as a warning sign."

The Independent gave several other examples of regions of potential conflict. These include Israel, Jordan and Palestine, who all rely on the Jordan River, which is controlled by Israel; Turkey and Syria, where Turkish plans to build dams on the Euphrates River brought the country to the brink of war with Syria in 1998, and where Syria now accuses Turkey of deliberately meddling with its water supply; China and India, where the Brahmaputra River has caused tension between the two countries in the past, and where China's proposal to divert the river is re-igniting the divisions; Angola, Botswana and Namibia, where disputes over the Okavango water basin that have flared in the past are now threatening to re-ignite as Namibia is proposing to build a three-hundred-kilometer pipeline that will drain the delta; Ethiopia and Egypt, where population growth is threatening conflict along the Nile; and Bangladesh and India, where flooding in the Ganges caused by melting glaciers in the Himalayas is wreaking havoc in Bangladesh, leading to a rise in illegal, and unpopular, migration to India.

While not likely to lead to armed conflict, stresses are growing along the U.S.-Canadian border over shared boundary waters. In particular, concerns are growing over the future of the Great Lakes, whose waters are becoming increasingly polluted and whose water tables are being steadily drawn down by the huge buildup of population and industry around the basin. A joint commission set up to oversee these waters was recently bypassed by the governors of the American states bordering the Great Lakes, who passed an amendment to the treaty governing the lakes that allows for water diversions to new communities off the basin on the American side. Canadian protests fell on deaf ears in Washington. In 2006, the U.S. government announced plans to have the U.S. coast guard patrol the Great Lakes using machine guns mounted on their vessels and revealed that it had created thirty-four permanent live-fire training zones along the Great Lakes from where it had already conducted a number of automatic weapons drills due to fierce opposition, firing three thousand lead bullets each time into the lakes. The Bush administration has temporarily called off these drills but is clearly asserting U.S. authority over what has in the past been considered joint waters.

Similar trouble is brewing on the U.S.-Mexican border, where a private group of U.S.–based water rights holders is using the North American Free Trade Agreement to challenge the long-term practice by Mexican farmers to divert water from the Rio Grande before it reaches the United States.

Water Refugees

Lester Brown of the Earth Policy Institute in Washington warns of a future of mass migration of water refugees. In his 2006 book, *Plan B 2.0: Rescuing a Planet under Stress and a Civilization in Trouble*, he presents a chilling picture of a not-so-distant future where water shortages have ended the growth in irrigated land and millions of displaced persons follow water sources for survival. He points out that one way of measuring water security is the amount of water available per person in a country. In 1995, 166 million people lived in countries where the average supply of water was less than 1,000 cubic meters of water a year – the amount deemed necessary to satisfy the basic needs of life. By 2050, Brown notes, 1.7 billion people will live in dire "water poverty" and be forced to relocate.

Already, he notes, water refugees can be found in Iran, Afghanistan, parts of Pakistan, in northwest China, and many parts of Africa. Right now, villages are being abandoned but eventually whole cities might have to be relocated, such as Sana'a, the capital of Yemen, or Quetta, the capital of Pakistan's Baluchistan Province. Chinese scientists report that there are now desert refugees in three provinces – Inner Mongolia, Ningxia and Gansu. Another four thousand villages face abandonment due to shrinking water supplies. In Iran, villages abandoned because of spreading deserts and lack of water number in the thousands. In Nigeria, thirty-five hundred square kilometers of land are converted to desert each year, making desertification the country's leading environmental problem. There, as elsewhere, farmers are forced to the outskirts of growing megacities' slums where they exacerbate the urban water crisis.

Every day, says Brown, bodies wash up on the shores of Italy, France and Spain: the result of desperate acts by desperate people fleeing drought. Every day, hundreds of Mexicans risk their lives trying to cross the U.S. border; some 400 to 600 Mexicans leave rural areas every day, abandoning plots of land too dry to farm. To get to the United States, they must pass across rivers so toxic (from waste dumping by foreign-controlled *maquiladora* plants), they have to wear plastic bags over their shoes.

Water Is Becoming a Global Security Issue

The United States

Water has recently (and suddenly) become a key strategic security and foreign policy priority for the United States. In the wake of the terrorist attacks of 9-11, protection of U.S. waterways and drinking water supplies from terrorist attack became vitally important to the White House. When Congress created the Department of Homeland Security in 2002, it gave the department responsibility for securing the nation's water infrastructure and allocated us$548 million in appropriations for security of water infrastructure facilities, funding that was increased in subsequent years. The Environmental Protection Agency created a National Homeland Security Research Center to develop the scientific foundations and tools to be used in the event of an attack on the nation's water systems, and a Water Security Division was established to train water utility personnel on security issues. It also created a Water Information Sharing and Analysis Center for dissemination of alerts about potential threats to drinking water and, with the American Water Works Association, a rapid e-mail notification system for professionals called the Water Security Channel. Ever true to market economy ideology, the Department of Homeland Security's mandate includes promoting public-private partnerships in protecting the nation's water security.

But the interest in water did not stop there. Water is becoming as important a strategic issue as energy in Washington. In an August 2004 briefing note for the Institute for the Analysis of Global Security, a think tank that focuses on the link between energy and security, Dr. Allan R. Hoffman, a senior analyst for the U.S. Department of Energy, declared that the energy security of the United States actually depends on the state of its water resources and warns of a growing water-security crisis worldwide. "Just as energy security became a national priority in the period following the Arab Oil Embargo of 1973–74, water security is destined to become a national and global priority in the decades ahead," says Hoffman. He notes that central to addressing water security issues is finding the energy to extract water from underground aquifers, transport water through pipelines and canals, manage and treat water for reuse and desalinate brackish and sea water – all technologies now being promoted by U.S. government partnerships with American companies. He also points out that the U.S. energy interests in the Middle East could be threatened by water conflicts in the region: "Water conflicts add to the instability of a region on which the U.S. depends heavily for oil. Continuation or inflammation of these conflicts could subject U.S. energy supplies to blackmail again, as occurred in the 1970s."

Water shortages and global warning pose a "serious threat" to America's national security, top retired military leaders told the president in an April 2007 report published by the national security think tank CNA Corporation. Six retired admirals and five retired generals warned of a future of rampant water wars into which the United States will be dragged. Erik Peterson, director of the Global Strategy Institute of the Center for Strategic and International Studies, a research organization in Washington that calls itself a "strategic planning partner for the government," says that the United States must make water a top priority in foreign policy. "There is a very, very critical dimension to all these global water problems here at home," he told *Voice of America News*. "The first is that it's in our national interest to see stability and security and economic development in key areas of the world, and water is a big factor with that whole set of challenges." His center has joined forces with ITT Industries, the giant water technology company; Proctor & Gamble, which has created a home water purifier called PUR and is working with the UN in a joint public-private venture in developing countries; Coca-Cola; and Sandia National Laboratories to launch a joint-research institute called Global Water Futures (GWF). Sandia, whose motto is "securing a peaceful and free world through technology" and that works to "maintain U.S. military and nuclear superiority," is contracted out to weapons manufacturer Lockheed Martin by the U.S. government, to operate, thus linking water security to military security in a direct way.

The mandate of Global Water Futures is twofold: to affect U.S. strategy and policy regarding the global water crisis and to develop the technology necessary to advance the solution. In a September 2005 report, Global Water Futures warned that the global water crisis is driving the world toward "a tipping point in human history," and elaborated on the need for the United States to start taking water security more seriously: "In light of the global trends in water, it is clear that water quality and water management will affect almost every major U.S. strategic priority in every key region of the world. Addressing the world's water needs will go well beyond humanitarian and economic development interests. . . . Policies focused on water in regions across the planet must be regarded as a critical element in U.S. national security strategy. Such policies should be part of a broader, comprehensive, and integrated U.S. strategy toward the global water challenges."

Innovations in policy and technology must be tightly linked, says the report, no doubt music to the ears of the corporations that sponsored it. GWF calls for closer innovation and cooperation between governments and the private sector and "redoubled" efforts to mobilize public-private partnerships in the development of technological solutions. And, in language that will be familiar to critics of the Bush administration who argue that the United States is not in Iraq to promote democracy, but rather to secure oil resources and make huge profits for American companies in the "rebuilding" effort, the report links upholding American values of democracy with the profit to be gained in the process: "Water issues are critical to U.S. national security and integral to upholding American values of humanitarianism and democratic development. Moreover, engagement with international water issues guarantees business opportunity for the U.S. private sector, which is well positioned to contribute to development and reap economic reward." Listed among the U.S. government agencies engaged in water issues in the report is the Department of Commerce, which "facilitates U.S. water businesses and market research, and improves U.S. competitiveness in the international water market."

Guarani Aquifer

There is growing concern in South America over the interest the United States is taking in the region's largest underground reservoir of water, the Guarani Aquifer, which stretches beneath parts of Argentina, Brazil, Paraguay and Uruguay – an area larger than Texas and California combined that now serves at least five hundred cities and towns in the region. *National Geographic News* reports that accusations are clouding international efforts to sustainably develop the Guarani because of the presence of a large U.S. army base in the area and the involvement of the Global Environment Facility, a U.S.–based funding consortium managed by the World Bank and the United Nations that involves U.S. private interests. The Brazilian civil society movement Grito Das Aguas has expressed concern that the United States will have access to the accumulated knowledge of years of research from Latin American universities to put at the disposal of American corporations. This "hydro-geopolitical" domination would simply be the latest chapter in a history of resource exploitation of the area, asserts Adolfo Esquivel, an Argentine activist and Nobel Peace Prize Laureate. (Unconfirmed reports have swirled for several years through the region that George W. Bush has bought 40,500 hectares (100,000 acres) of ranchland in northern Paraguay right on the aquifer. *The Guardian* newspaper quotes Erasmo Rodriguez Acosta, the governor of the Alto Paraguay region where the new acquisition supposedly lies, as confirming this deal. These rumors have done nothing to dispel the growing concern about U.S. interests in the Guarani.)

Although the project is widely referred to as an environmental and humanitarian effort, local groups have reason to be concerned that private investors will be brought in at some point. On its website, the Global Environment Facility openly touts the private sector as a partner in its projects: "It is abundantly clear that global environmental problems like climate change and biodiversity loss will be solved only if the private sector also weighs in with its vast technical, managerial and financial resources and expertise. . . . The private sector is recognized as an important stakeholder in GEF activities and has a critical role to play in addressing global environmental challenges in partnership with the GEF." The website states that the World Bank plays the primary role in ensuring and managing private sector involvement in GEF projects and notes that more than twelve climate change projects funded by the GEF involve the participation of energy service companies. It adds, "Brazil's Ministry of the Environment has been actively discussing mechanisms to ensure that private sector businesses and business foundations with an interest in biodiversity protection can participate in GEF-funded project activities."

Europe

France, Germany and Great Britain have long promoted the interests of their water corporations in the global South, going so far as to involve their embassies in negotiations between countries in need of services and their domestic companies. France in particular has been an aggressive negotiator in developing countries, using its clout, as well as foreign and aid policy, to get the best deals possible for Suez and Veolia. Now, a new European organization with remarkable similarities to Global Water Futures has been founded to examine the issue of water security for Europe.

The European Water Partnership (EWP) was established in 2006 to "find solutions for the water challenges in wider Europe" and to stimulate partnerships between governments, civil society and the private sector. It will undertake "a strategic research agenda" and act as "vital mechanism to increase investment and support toward the competitiveness of the European water industry." The EWP came out of two major water initiatives, the Water Supply and Sanitation Technology Platform and the European Regional Process, which produced the European position at the Fourth World Water Forum in Mexico City. The Water Supply and Sanitation Technology Platform is a European Commission–backed "technology platform" to promote technological innovation in water and enhance the competitiveness of the European water industry. Its board chairman is Dr. Claude Roulet, vice-president of Texas-based Schlumberger, a giant energy technology company operating in eighty countries. Roulet is also vice-president of EUROGIF, the European Oil and Gas Innovative Forum, a 2,500-member strong industry group created to promote the competitive position of Europe's energy sector.

Many members of the European Water Partnership are water technology companies, clearly demonstrating that, like Global Water Futures in the United States, this project makes a connection between the security of Europe's water and the promotion of Europe's private sector water reuse industry. Other members are water technology industry associations, universities, government research institutes and even legal and consulting firms who specialize in advising water companies. Each June, the EWP holds the European Policy Summit on Water, a major policy summit "to focus on water security and what we have to do collectively and corporately." The summit is co-sponsored by the Friends of Europe, a prominent Brussels-based think tank whose members included a who's who of senior retired politicians and former members of the trilateral commission. The president of Friends of Europe is Viscount Etienne Davignon, vice-president of Suez-Tractebel, a wholly owned engineering subsidiary of Suez, and trustees include Pascal Lamy, director-general of the World Trade Organization, and Joachim Bitterlich, executive vice-president of Veolia. (Davignon also advises European Development Commissioner Louis Michel on aid polices for Africa, an appointment that has come under intense criticism from civil society groups because of Davignon's Suez connection.)

The 2007 EWP summit was called "Water Security – Does Europe Have a Strategy?" and sessions, which included "Enhancing water and energy security" and "Investment opportunities in water," clearly made the same link as Washington has between water security and advancing the interests of the water technology sector. The EWP has also launched an interactive website and blog called "Blue Gold" (which amused this author, who first coined the term as a critique of those who see water as a profit-making resource), with posted entries such as "Bottled water benefits the poor" and "European Commission aims to adopt market-based instruments to achieve environmental goals."

China

Water as a matter of national security has also moved up to the highest levels of the Chinese government, and here, as in Europe and the United States, the answer lies in private sector technology innovation combined with national security policy. The major difference is that China does not yet have its own private water technology industry and must rely on the established industries of Europe and the United States. China's Ministry of Water Resources, a huge bureaucracy spanning ten departments, is responsible for all aspects of water policy, planning and financing, and the oversight of "economic regulatory measures," including the promotion of a massive water technology infrastructure. "Water has become a hot issue for the Chinese government and society," declares the ministry on its official website. Every year, under the direction of the Chinese government, it hosts a Water Expo in Beijing, which showcases the "latest products and water technologies" and has become "a brand-new platform serving the Chinese water resources industry."

The ministry also endorsed the Beijing April 2007 China Water Congress, whose mandate is "to ensure the great demand of water supply and treatment in China into a profitable age." Co-sponsored by the International Desalination Association and Motimo, a Chinese membrane technology company, the congress brought together Chinese officials and the global water industry, including Suez, Veolia and hundreds of water technology corporations, such as ITT and Hyflux. "China water is a huge market waiting for the whole world to participate in," said Qiu Baoxing, China's minister of construction. "Current trends in investment and policymaking are driving water projects inexorably toward privatization. The industry is becoming more open than ever."

At the same time, the Chinese government (like the U.S. government) is working to secure new sources of water outside China's borders. Tibet is the source of water for close to half of humanity as the ten major watersheds that form high above the Tibetan Plateau spread out all across Asia. China's plan to permanently divert seventeen cubic billion meters a year from these watersheds is creating great consternation with Tibetans (who consider Tibet an occupied country) as well as the other Asian countries who depend on that water for daily survival. Tashi Tsering, a Tibetan expert on natural resources at the University of British Columbia, told Radio Free Europe that this project mirrors a pattern of exploitation of Tibetan natural resources: "They look up to Tibet for water supplies, which is perfect for these huge water-construction bureaucracies and businesses basically who are looking to see where else they can build dams and water-diversion projects because they have already diverted and dammed all the rivers in China."

These power blocs are setting the stage to secure control over water supplies in order to protect and promote their own national security interests. As well, they want to maintain a competitive edge in an increasingly competitive global economy and for this, they need water. The United States and Europe are also seeking to promote the interests of their water corporations in the race to create a global water cartel. For China, securing supplies is a matter of life and death as an emerging economic superpower. As commodities analyst Jim Rogers told the *Hong Kong Standard*, "If China cannot solve the water problem, that could be the end of the story." These superpower dynamics add a whole new dimension to the growing concerns around the potential for conflict and wars over water.

Blue Covenant: The Alternative Water Future

Humanity still has a chance to head off these scenarios of conflict and war. We could start with a global covenant on water. The Blue Covenant should have three components: a *water conservation* covenant from people and their governments that recognizes the right of the Earth and of other species to clean water, and pledges to protect and conserve the world's water supplies; a *water justice* covenant between those in the global North who have water and resources and those in the global South who do not, to work in solidarity for water justice, water for all and local control of water; and a *water democracy* covenant among all governments acknowledging that water is a fundamental human right for all. Therefore, governments are required not only to provide clean water to their citizens as a public service, but they must also recognize that citizens of other countries have the right to water as well and to find peaceful solutions to water disputes between states.

A good example of this is the Good Water Makes Good Neighbors project of Friends of the Earth Middle East, which seeks to use shared water and the notion of water justice to negotiate a wider peace accord in the region. Another example is the successful restoration of the beautiful Lake Constance by Germany, Austria, Lichtenstein and Switzerland, the four countries that share it.

The Blue Covenant should also form the heart of a new covenant on the right to water to be adopted both in nation-state constitutions and in international law at the United Nations. To create the conditions for this covenant will require a concerted and collective international collaboration and will have to tackle all three water crises together with the following alternatives.

Water Conservation

The alternative to crisis one – dwindling freshwater supplies – is conservation. A great deal of work has been done to document the ways in which we can save the planet's water systems. The knowledge and recommendations are there; what we lack is political will. The first and most important step is the restoration of watersheds and the protection of water sources. Slovakian scientist Michal Kravçik and his colleagues believe that our collective abuse of water is the most important factor in climate change and warn that with time, our current behavior will completely destroy the hydrologic cycle. They argue that the only solution is the massive restoration of watersheds. Bring water back into parched landscapes, they argue. Return water that has disappeared by retaining as much rainwater as possible within the country so that water can permeate the soil, replenish groundwater systems and return to the atmosphere to regulate temperatures and renew the hydrologic cycle. All human, industrial and agricultural activity must conform to this imperative, a project that could also employ millions and alleviate poverty in the global South. Our cities must be ringed with green conservation zones and we must restore forests and wetlands – the lungs and kidneys of freshwater.

Three basic laws of nature must be addressed. First, it is necessary to create the conditions that allow rainwater to remain in local watersheds. This means restoring the natural spaces where rainwater can fall and where water can flow. Water retention can be carried out at all levels: roof gardens in family homes and office buildings; urban planning that allows rainwater to be captured and returned to the earth; water harvesting in food production;

capturing daily water discharge and returning it clean to the land, not to the rising oceans.

Second, we cannot continue to mine groundwater supplies at a rate greater than natural recharge. If we do, there will not be enough water for the next generation. Extractions cannot exceed recharge just as a bank account cannot be drawn down without new deposits. Governments everywhere must undertake intensive research into their groundwater supplies and regulate ground-water takings before their underground reservoirs are gone.

Third, of course, we must stop polluting our surface and groundwater sources and we must back up this intention with strict legislation. Martin Luther King Jr. said, "It may be true that the law cannot change the heart but it can restrain the heartless." Legislation should also include penalties for domestic corpora-tions that pollute in foreign countries. Water abuse in oil and methane gas production must stop. Much has been written on the harm to water of industrial and chemical-based agriculture and flood irrigation. In their 2007 book, *Who Owns the Water?*, editors Klaus Lanz, Christian Rentsch, Rene Schwarzenbach and Lars Müller call for a Blue Revolution in agriculture, getting "more crop per drop" and a cessation of the mass use of chemicals to grow food. They point out that today, farmers around the world use six times more pesticides than they did fifty years ago. We must listen to the many voices sounding the alarm around the rush toward water-guzzling biofuel farming, a new and dangerous use of productive farmland, heavily subsidized by many governments. Sandra Postel and others point the way to more sustainable food production systems including the use of drip irrigation.

The International Forum on Globalization has written extensively on the notion of "subsidiarity," whereby nation-state policies and international trade rules could support local food production in order to protect the environment and promote local sustainable agriculture. Such policies would also discourage the virtual trade in water, and countries could ban or limit the mass movement of water by pipeline. Government investment in water and wastewater infrastructure would save huge volumes of water lost every day in old or nonexistent systems. Domestic laws could enforce water-harvesting practices at every level.

Kravçik is no wide-eyed idealist. He knows that this nature-based solution challenges the deepest tenets of economic globalization and the growth imperative behind it. (The late American environmentalist Edward Abbey said that growth for the sake of growth is the ideology of the cancer cell.) Kravçik also knows that this plan would undermine the massive investment now going into technological solutions such as desalination, wastewater reuse and nanotechnology. "The tragedy of our solution," he writes, "is that it is not a magnificent and attractive technical-engineering business for big companies to invest in, but rather a community program aimed at meticulous care of thousands of people." He calls on governments and international institutions to save the Blue Planet by means of "community sustainable development programs" that would be many times cheaper than the technology they are supporting and protect biodiversity and prevent natural disasters and wars.

Many examples abound, from the New Mexican "Acequia" system that uses an ancient natural ditch irrigation tradition to distribute water in arid lands to Rajendra Singh's water harvesting project in India. The International Rainwater Harvesting Alliance based in Geneva works globally to promote sustainable rainwater harvesting programs and government and United Nations support for them. Kravçik says we have ten years to implement these reforms. Simply put, the water in the hydrologic cycle will provide for us forever if we care for it and allow the earth to renew it.

Water Justice
The alternative to crisis two – inequitable access to water – is water justice, not charity. Millions of people live in countries that cannot provide clean water (or healthcare or education) to their citizens as they are burdened by their debt to the World Bank and the International Monetary Fund. As a result, poor countries are forced to exploit both their people and their resources, like water,

to pay their debt. Groups and networks such as Jubilee South, Make Poverty History and ActionAid report that at least sixty-two countries need deep debt relief if the daily deaths of thousands of children are to end. As well, foreign aid in many northern countries falls far below the recommended 0.7 percent of GDP. The United States, for instance, spends only 0.17 percent of its GDP on overseas aid, and under the Bush administration, conditions that aid on the promise of open markets for American corporations.

What many of these corporations are doing in the global South is criminal, imposing a new form of colonial conquest dressed up as the one and only economic model available. In many countries, North American and European companies receive multiyear tax breaks and treat both the population and the local environment with contempt. As Dr. Dale Wen with the International Forum on Globalization explains, one cannot just condemn "China's pollution problem" without condemning the foreign transnationals causing so much of the damage on Chinese soil. First World governments need to take control of their corporate nationals in foreign countries. Canadian mining companies, for instance, are known environmental abusers and should be held to account for their actions by the Canadian government.

The water companies are among the worst and should be forced to leave poor countries. If the World Bank, the United Nations and northern countries were serious about providing clean water for all, they would cancel or deeply cut the Third World debt, substantively increase foreign aid, fund public services, tell their big bottling companies to stop draining poor countries dry and invest in water reclamation programs to protect source water. They would also tell the water companies that they no longer have any say in which countries and communities receive water funding. Citizens of First World countries need to recognize and challenge the hypocrisy of their governments, many of whom would never permit foreign corporations to run and profit from their water supplies, but who continue to support the global financial and trade institutions that commodify water in the Third World. Many in the water justice movement work with fair trade groups to create a whole new set of rules for global trade based on sustainability, cooperation, environmental stewardship and fair labor standards. They also promote a tax on financial speculation; even a modest tax could pay for every public hospital, school and water utility in the global south.

Special mention must be made of two groups feeling the brunt of water inequity: women and indigenous people. The Women's Environment and Development Organization (WEDO), an international advocacy organization that seeks to increase the power of women worldwide as policymakers, reminds us that women carry out 80 percent of water-related work throughout the world and therefore bear the brunt of water inequity. Water is a critical component of gender equality and women's empowerment, along with environmental security and poverty eradication, asserts WEDO. The more policy-making about water is moved from local communities to a global level (the World Bank, for instance), the less power women have to determine who gets it and under what circumstances. As the primary collectors of water throughout the world, women must be recognized as major stakeholders in the decision-making process.

Indigenous people are particularly vulnerable to water theft and appropriation, and their proprietary rights to their land and water must be protected by governments. In a call to action on International Water Day 2007 called *Honor the Water, Respect the Water, Be Thankful for the Water, Protect the Water*, the Indigenous Environmental Network (IEN) points out that many of the resources being plundered by governments and corporations of the global North lie on ancestral lands. The ensuing exploitation, privatization and contamination upsets the balance of cultural resources and sacred sites, says IEN, which has issued a challenge to "raise the Indigenous voice in defence of Sacred Water."

Water Democracy

The alternative to crisis three – the corporate control of water – is public control. The creation of a worldwide water cartel is wrong ethically, environmentally and socially and ensures that the decisions regarding the allocation of water are made based on commercial, not environmental or social, concerns. Private transnational corporations cannot maintain a competitive position in the water industry if they operate on the principles of water conservation, water justice and water democracy. Only governments, with their mandate to work in the public good, can operate on these principles. Water corporations, be they utilities, bottled water companies or the new water reuse industry, are dependent on increased consumption to generate profits and will never be able to seriously join the effort to protect and conserve source water. Further, the control of water supplies by corporations, usually foreign, dramatically reduces the democratic oversight of the communities and countries in which they operate. Water must be understood to be part of the global commons but clearly subject to local, democratic and public management. There are many alternatives to the corporate control of water and countless examples of where it is working.

Public Services International and the World Development Movement have done a great deal of work on alternatives to private control of water services and advocate public-public partnerships (PUPs). As David Hall and Emanuele Lobina explain in *Water as a Public Service*, water utilities have to have political, public legitimacy, legal powers, financial resources and a sustainable labor force. Established water operators in both the North and South have developed these capacities. But many in the South have not been able to do so yet. PUPs are a mechanism for providing capacity building for these countries, either through Water Operator Partnerships, whereby established public systems transfer expertise and skills to those in need, or through projects whereby public institutions such as public sector unions or public pension fund boards, use their resources to support public water services in developing countries. The objective is to provide local management and workers with the necessary skills to deliver water and provide wastewater services to the public.

Examples of successful PUPs include partnerships between Stockholm and Helsinki water authorities and the former Soviet

Union countries of Estonia, Latvia and Lithuania and between Amsterdam Water and cities in Indonesia and Egypt. PSI asserts that, if each effectively functioning public water utility in the world were to "adopt" just three cities in need, public-private partnerships could operate on a global basis, and provide water to all those in need at a fraction of the cost now encountered supporting the private companies. This would also become a concrete example of how cooperation over water could be a uniting force for humanity.

In its March 2007 publication, *Going Public: Southern Solutions to the Global Water Crisis*, the World Development Movement gives examples of four successful local public water systems in Porto Alegre, Brazil; Tamil Nadu, India; Phnom Penh, Cambodia; and Kampala, Uganda. It finds that where all are different and provide local solutions to local problems, all have in common a commitment to efficiency, accountability, transparency and community participation. PSI has also written extensively on financing public water and recommends a combination of progressive central government taxation, micro-financing and cooperatives to run the systems on a day-to-day basis. To finance capital investments, PSI recommends borrowing from public sector national and international sources to shield countries and investors from currency risks. Development banks should get back to the role they were created for and invest in efficient, accountable, transparent and democratic not-for-profit public systems.

Corporate control of water in other areas must be confronted as well. That is not to say there is no role for the private sector in finding solutions to the global water crisis. But all private sector activity must come under strict public oversight and government accountability, and all would have to operate within a program whose goals are conservation and water justice. There would be a very different role for the private water reuse technology industry if the world were to adopt Michal Kravçik's rainwater harvesting solution as well as strict anti-pollution and source protection laws. The future would not include thousands of desalination plants ringing the world's oceans or machines sucking rainwater out of the clouds. Nor would there be a reason, real or perceived, to drink bottled water.

But governments cannot wait for these changes to implement strict controls over the water reuse technology industry and all government investment in this sector must clearly be geared toward the public good. Similarly, in countries or communities where bottled water is still the only safe water to drink, governments must control the bottling industry, insisting that it be sustainable, locally run and publicly controlled, and the bottles themselves recyclable. The end goal, however, must be to do away with the need for bottled water everywhere.

The Right to Water: An Idea Whose Time Has Come

Finally, the global water justice movement is demanding a change in international law to settle once and for all the question of who controls water. It must be commonly understood that water is not a commercial good, although of course it has an economic dimension, but rather a human right and a public trust. What is needed now is binding law to codify that states have the obligation to deliver sufficient, safe, accessible and affordable water to their citizens as a public service. While "water for all, everywhere and always" may appear to be self-evident, the fact is that the powers moving in to take corporate control of water have resisted this notion fiercely. So have many governments, either because, in the case of rich governments, their corporations benefit from the commodification of water or, in the case of poor governments, because they fear they would not be able to honor this commitment. So groups around the world are mobilizing in their communities and countries for constitutional recognition of the right to water within their borders and at the United Nations for a full treaty that recognizes the right to water internationally. (The terms *covenant*, *treaty* and *convention* are used interchangeably at the UN.)

Rosmarie Bar of Switzerland's Alliance Sud explains that behind the call for a binding convention or covenant are questions of principle. Is access to water a human right or just a need? Is water a common good like air or a commodity like Coca-Cola? Who is being given the right or the power to turn the tap on or off – people, governments or the invisible hand of the market? Who sets the price for a poor district in Manila or La Paz – the locally elected water board or the CEO of Suez? The global water crisis cries out for good governance, says Bar, and good governance needs binding, legal bases that rest on universally applicable human rights. A UN covenant would set the framework for water as a social and cultural asset, not an economic commodity. As well, it would establish the indispensable legal groundwork for a just system of distribution. It would serve as a common, coherent body of rules for all nations, rich and poor, and clarify that it is the role of the state to provide clean, affordable water to all of its citizens. Such a covenant would also safeguard already accepted human rights and environmental principles in other treaties and conventions.

Michigan lawyer Jim Olson, who has been deeply involved in the fight against Nestlé, says the point must be "repeated and repeated" that privatization of water is simply incompatible with the nature of water as a commons and therefore, with fundamental human rights. "Water is always moving unless there is human intervention. Intervention is the right to use, not own and privatize to the exclusion of others who enjoy equal access to use water. It is important to distinguish between sovereign ownership and control of water, enjoyed by states or nations through which water flows or moves, and private ownership. Sovereign state ownership is not the same and has to do with control and use of water for the public welfare, health and safety, not for private profit." If on the other hand, says Olson, the state sides with the World Bank and negotiates private rights to its water with corporations, that state has violated the rights of its citizens who would have redress under the principle of human rights if the covenant is well crafted.

A human rights convention or covenant imposes three obligations on states: the Obligation to Respect, whereby the state must refrain from any action or policy that interferes with the enjoyment of the human right; the Obligation to Protect, whereby the state is obliged to prevent third parties from interfering with the enjoyment of the human right; and the Obligation to Fulfill, whereby the state is required to adopt any additional measures directed toward the realization of that right. The Obligation to Protect would oblige governments to adopt measures restraining corporations from denying equal access to water (in itself an incentive for water companies to leave) as well as polluting water sources or unsustainably extracting water resources.

At a practical level, a right-to-water covenant would give citizens a tool to hold their governments accountable in their domestic courts and the "court" of public opinion, and for seeking international redress. Says the World Conservation Union, "Human rights are formulated in terms of individuals, not in terms of rights and obligations of states vis-à-vis other states as international law provisions generally do. Thus by making water a human right, it could not be taken away from the people. Through a rights-based approach, victims of water pollution and people deprived of necessary water for meeting their basic needs are provided with access to remedies. In contrast to other systems of international law, the human rights system affords access to individuals and NGOs."

The union also states that a right-to-water covenant would make both state obligations and violations more visible to citizens. Within a year of ratification, states would be expected to put in place a plan of action, with targets, policies, indicators and timeframes to achieve the realization of this right. As well, states would have to amend domestic law to comply with the new rights. In some cases, this will include constitutional amendments. Some form of monitoring of the new rights would also be established and the needs of marginalized groups such as women and indigenous peoples would be particularly addressed.

A covenant would also include specific principles to ensure civil society involvement to convert the UN convention into national law and national action plans. This would give citizens an additional constitutional tool in their fight for water. As stated in a 2003 manifesto on the right to water by Friends of the Earth Paraguay, "An inseparable part of the right is control and sovereignty of local communities over their natural heritage and therefore over the management of their sources of water and over the use of the territories producing this water, the watersheds and aquifer recharge areas." A right-to-water covenant would also set principles and priorities for water use in a world destroying its water heritage. The covenant we envisage would include language to protect water rights for the Earth and other species and would address the urgent need for reclamation of polluted waters and an end to practices destructive of the world's water sources. As Friends of the Earth Paraguay put it, "The very mention of this supposed conflict, water for human use versus water for nature, reflects a lack of consciousness of the essential fact that the very existence of water depends on the sustainable management and conservation of ecosystems."

Progress at the United Nations

Water was not included in the 1947 United Nations Universal Declaration of Human Rights because at that time water was not perceived to have a human rights dimension. The fact that water is not now an enforceable human right has allowed decision-making over water policy to shift from the UN and governments toward institutions and organizations that favor the private water companies and the commodification of water such as the World Bank, the World Water Council and the World Trade Organization. However, for more than a decade, calls have been made at various levels of the United Nations for a right-to-water convention. Civil society groups argued that, because the operations of the water companies had gone global and were being backed by global financial institutions, nation-state instruments to deal with water rights were no longer sufficient to protect citizens. International laws were needed, we argued, to control the global reach of the water barons. We also noted that at the 1990 Rio Earth Summit, the key areas of water, climate change, biodiversity and desertification were all targeted for action. Since then, all but water have resulted in a UN convention.

This lobbying started to pay off and the right to water was recognized in a number of important international UN resolutions and declarations. These include the 2000 General Assembly Resolution on the Right to Development; the 2004 Committee on Human Rights resolution on toxic wastes; and the May 2005 statement by the 116-member Non-Aligned Movement on the right to water for all. Most important is General Comment Number 15, adopted in 2002 by the UN Committee on Economic, Social and Cultural Rights, which recognized that the right to water is a prerequisite for realizing all other human rights and "indispensable for leading a life in dignity." (A General Comment is an authoritative interpretation of a human rights treaty or convention by an independent committee of experts that has a mandate to provide states with an interpretation of the treaty or convention. In this case, the interpretation applies to the International Covenant on Economic, Social and Cultural Rights.) General Comment Number 15 is therefore an authoritative interpretation that water is a right and an important milestone on the road to a full binding UN convention.

But as John Scanlon, Angela Cassar and Noemi Nemes of the World Conservation Union point out in their 2004 legal briefing paper, *Water as a Human Right?* General Comment Number 15 is an interpretation, not a binding treaty or convention. To clearly bind the right to water in international law, a binding covenant is needed. So the pressure for a full covenant intensified. In early 2004, Danuta Sacher of Germany's Bread for the World and Ashfaq Khalfan of the Right to Water program at the UN Center on Housing Rights and Evictions called a summit and a new international network called Friends of the Right to Water was born. The network set out to mobilize other water justice groups and national governments to join the campaign to strengthen the rights established in General Comment Number 15 and put in place the mechanisms to ensure implementation of the right to water through a covenant.

In November 2006, responding to a call from several countries, the newly formed UN Human Rights Council requested the
Office of the High Commissioner for Human Rights to conduct a
detailed study on the scope and content of the relevant human
rights obligations related to access to water under international
human rights instruments, and to include recommendations for
future action. While the request does not specifically refer to a
covenant, many see this process as having the potential to lead to
one. In April 2007, Anil Naidoo of the Council of Canadians' Blue
Planet Project, another founding member of Friends of the Right
to Water, organized to present a letter of endorsement calling for
a right-to-water covenant to UN Commissioner Madam Louise
Arbour, signed by 176 groups from all over the world.

It has been essential to gain the support of governments in the
global South, many of whom fear that their citizens could use a
covenant against them if they are unable to immediately fulfill their
new obligation. Proponents of a covenant emphasize that the appli-
cation of a new human rights obligation is understood to be
progressive. States without the power to implement the full right
are not held accountable for not immediately delivering. What is
required is the need to rapidly take minimal steps for implementation that will increase as capacity increases. But some governments
are using their incapacity as an excuse to cover real priorities, such
as funding the military rather than public services. A rights-based
approach to development distinguishes between inability and
unwillingness. As agreed at the 1993 UN World Conference on
Human Rights, "While development facilitates the enjoyment of
all human rights, the lack of development may not be invoked to
justify the abridgement of internationally recognized human
rights." A government that fails to ratify a right-to-water covenant
should not try to hide behind capacity arguments.

Nor should relatively water-rich governments such as
Canada hide behind a false fear (which Canada is doing) that they
will be forced to share the actual water supplies within their
territories. A human rights treaty is between a nation-state and its
citizens. Recognition of the right to water in no way affects a

country's sovereign right to manage its own water resources. What will be expected of First World governments and their development agencies is adequate aid to help developing countries meet their goals and ensure that their aid, and that of the World Bank, is directed toward not-for-profit public water services.

Dueling Visions

While the global water justice movement is excited and encouraged by these developments, there is a growing concern that this process could be hijacked by the water corporations, some northern countries and the World Bank, and used to create a convention that would enshrine the inclusion of private sector players. There is now a widespread understanding that the right to water is an idea whose time has come and some who opposed it until very recently have decided to drop their opposition and help shape both the process and the end product in their image. The irony here is that this new scenario may just have arisen out of the very success of the global water justice movement's hard work. Until recently, the global institutions and the big water companies adamantly opposed a right to water convention. So did many European countries such as France, England and Germany, home to the big water companies. At the World Water Forums in The Hague and Kyoto, World Water Council members and governments refused civil society calls for a right-to-water convention and said that water is a human need, not a human right. These are not semantics: you cannot trade or sell a human right or deny it to someone on the basis of inability to pay.

At the Fourth World Water Forum in Mexico City, the ministerial declaration once again did not include the right to water. But the World Water Council did release a new report called *The Right to Water: From Concept to Implementation*, a bland restatement of many UN documents with almost no mention of the private sector (except to say that the right to water can be implemented in a "variety of ways") and with no reference to the public-private debate raging around it. While the report falls far short of recommending a convention on the right to water, the first words of the foreword (written by Loïc Fauchon, president of the World Water Council and senior executive with Suez) capture the essence of the situation in which these corporations and the World Bank now find themselves: "The right to water is an element that is indissociable from human dignity. Who, today, would dare say otherwise?" Who indeed?

The World Water Council is working with Green Cross International, an environmental education organization headed up by Mikhail Gorbachev, which has launched its own high-profile campaign for a UN convention on the right to water, and it is just the sort of convention that Loïc Fauchon could live with. Although the Green Cross draft convention admits that there is a problem with "excessive profits and speculative purposes" in the private exploitation of water, it nevertheless places the commercial and human right to water on an equal footing, sets the stage for private financing for water services, allows for the private management of water utilities and says that water systems should follow market rules. In a legal analysis of the Green Cross draft convention, Steven Shrybman, a Canadian trade expert and legal counsel to Canada's Blue Planet Project, says it is "so seriously flawed as to represent a retreat from current international legal protection for the human right to water." Yet Gorbachev defended his pro-corporate proposal in an interview with the *Financial Times* when he said that corporations are the "only institutions" with the intellectual and financial potential to solve the world's water problems and that he is "prepared to work with them."

The global water justice movement would never endorse a convention or covenant of this kind. In submissions to the High Commissioner, hundreds of groups have urged the United Nations to take a clear stand in favor of public ownership of water. For them, a covenant must explicitly describe water not only as a human right but also as a public trust. As well, a UN covenant on the right to water will have to address the two great shortcomings of existing human rights law if it is to be accepted by civil society. Those shortcomings are their failure to establish meaningful enforcement mechanisms and the failure to bind international bodies.

In his submission to Madam Arbour, lawyer Steve Shrybman said that the most significant development in international law has not been taking place under the auspices of the United Nations, but rather, under the World Trade Organization and the thousands of bilateral investment treaties between governments that have codified corporate rights into international law. "Under these rules, water is regarded as a good, an investment and service, and as such, it is subject to binding disciplines that severely constrain the capacity of governments to establish or maintain policies, laws and practices needed to protect human rights, the environment or other non-commercial societal goals that may impede the private rights entrenched by these trade and investment agreements."

Moreover, states Shrybman, these agreements have equipped corporations with powerful new tools in asserting proprietary rights over water with which the state cannot interfere. "The codification of such private rights creates an obvious and serious impediment to the realization of the human right to water." Private tribunals operating under these treaties are now engaged in arbitrating conflicts between human rights norms and those of investment and trade law – a role they are ill-suited to serve. He goes on to challenge the High Commissioner to recognize the need to deal with this reality and warns that unless UN bodies are able to reassert their role as the fundamental arbiter of human rights, they risk becoming bystanders as private tribunals operating entirely outside the UN framework resolve key questions of human rights law. To be effective, the covenant must assert the primacy of the human right to water where there is a conflict with private and commercial interests. As well, this instrument must apply to other institutions beside states, most importantly, transnational corporations, the WTO and the World Bank.

Grassroots Take the Lead

Clearly, the stage has been set for another form of contest. Having been successful in forcing the United Nations to deal with the right to water, the global water justice movement must now work hard to ensure it is the right kind of instrument. There are many good signs. While several important countries remain opposed to the right to water, most notably the United States, Canada, Australia and China, many more have come on board in recent years. The European Parliament adopted a resolution acknowledging the right to water in March 2006, and in November 2006, as a response to the 2006 UN Human Development Report on the world's water crisis, Great Britain reversed its opposition and recognized the right to water. As Ashfaq Khalfan of the Centre on Housing Rights and Evictions explains, most countries in one form or another have supported the notion of the right to water in various resolutions at the United Nations and can be counted on to do so again. The challenge is to get support for a covenant that will truly be able to deliver on the promise. This is where civil society groups can be so effective. In many countries, water justice groups are hard at work to convince their governments to support the right kind of tool.

But they are not waiting for the United Nations. Many are also working hard within their countries to assert the right of water for all through domestic legislative changes. On October 31, 2004, the citizens of Uruguay became the first in the world to vote for the right to water. Led by Adriana Marquisio and Maria Selva Ortiz of the National Commission for the Defence of Water and Life and Alberto Villarreal of Friends of the Earth Uruguay, the groups first had to obtain almost three hundred thousand signatures on a plebiscite (which they delivered to Parliament as a "human river"), in order to get a referendum placed on the ballot of the national election calling for a constitutional amendment on the right to water. They won the vote by an almost two-thirds majority, an extraordinary feat considering the fear-mongering that opponents mounted. The language of the amendment is very important. Not only is water now a fundamental human right in Uruguay, but also social considerations must now take precedence over economic considerations when the government makes water policy. As well, the constitution now reflects that "the public service of water supply for human consumption will be served exclusively and directly by state legal persons" that is to say, not by corporations.

Several other countries have also passed right-to-water legislation. When apartheid was defeated in South Africa, Nelson Mandela created a new constitution that defined water as a human right. However, the amendment was silent on the issue of delivery and soon after, the World Bank convinced the new government to privatize many of its water services. Several other developing countries such as Ecuador, Ethiopia and Kenya also have references in their constitutions that describe water as a human right, but they, too, do not specify the need for public delivery. The Belgium Parliament passed a resolution in April 2005 seeking a constitutional amendment to recognize water as a human right, and in September 2006, the French Senate adopted an amendment to its water bill that says that each person has the right to access to clean water. But neither country makes reference to delivery. The only other country besides Uruguay to specify in its constitution that water must be publicly delivered is Netherlands, which passed a law in 2003 restricting the delivery of drinking water to utilities that are entirely public. But Netherlands did not affirm the right to water in this amendment. Only the Uruguayan constitutional amendment guarantees both the right to water and the need to deliver it publicly and is therefore, a model for other countries. Suez was forced to leave the country as a direct result of this amendment.

Other exciting initiatives are underway. In August 2006, the Indian Supreme Court ruled that protection of natural lakes and ponds is akin to honoring the right to life – the most fundamental right of all according to the court. Activists in Nepal are going before their Supreme Court arguing that the right to health guaranteed in the country's constitution must include the right to water. The Coalition in Defense of Public Water in Ecuador is celebrating its victory over the privatization of its water by demanding that the government take the next step and amend the constitution to recognize the right to water. The Coalition

Against Water Privatization in South Africa is challenging the practice of water metering before the Johannesburg High Court on the basis that it violates the human rights of Soweto's citizens. President Evo Morales of Bolivia has called for a "South American convention for human rights and access for all living beings to water" that would reject the market model imposed in trade agreements. At least a dozen countries have reacted positively to this call. Civil society groups are hard at work in many other countries to introduce constitutional amendments similar to that of Uruguay. Ecofondo, a network of sixty groups in Colombia, has launched a plebiscite toward a constitutional amendment similar to the Uruguayan amendment. They need at least one and a half million signatures and face several court cases and a dangerous and hostile opposition. Dozens of groups in Mexico have joined COMDA, the Coalition of Mexican Organizations for the Right to Water, in a national campaign for a Uruguayan-type constitutional guarantee to the right to water.

A large network of human rights, development, faith-based, labor and environmental groups in Canada has formed Canadian Friends of the Right to Water, led by the Blue Planet Project, to get the Canadian government to change its opposition to a UN covenant on the right to water. A network in the United States led by Food and Water Watch is calling for both a national water trust to ensure safekeeping of the nation's water assets and a change of government policy on the right to water. Riccardo Petrella has led a movement in Italy to recognize the right to water, which has great support among politicians at every level. Momentum is growing everywhere for a right whose time has come.

꧁꧂

This, then, is the task: nothing less than reclaiming water as a commons for the Earth and all people that must be wisely and sustainably shared and managed if we are to survive. This will not happen unless we are prepared to reject the basic tenets of market-based globalization. The current imperatives of competition, unlimited growth and private ownership when it comes to water must be replaced by new imperatives – those of cooperation, sustainability and public stewardship. As Bolivia's Evo Morales explained in his October 2006 proposal to the heads of states of South America, "Our goal needs to be to forge a real integration to 'live well.' We say 'live well,' because we do not aspire to live better than others. We do not believe in the line of progress and unlimited development at the cost of others and nature. 'Live well' is to think not only in terms of income per capita but cultural identity, community, harmony between ourselves and with mother earth."

There are lessons to be learned from water, nature's gift to humanity, which can teach us how to live in harmony with the earth and in peace with one another. In Africa, they say, "We don't go to water ponds merely to capture water, but because friends and dreams are there to meet us."

Sources and Further Reading

Chapter 1: Where Has All the Water Gone?

This chapter deals with the ecological and human world water crisis, which has been thoroughly documented in recent years. There is much source material for this chapter.

The United Nations monitors the global water crisis in a number of its agencies. Through its World Water Assessment Programme, which coordinates the work of twenty-four agencies, every three years the UN publishes a groundbreaking assessment of the world's water. Its 2006 World Water Development Report is called *Water: A Shared Responsibility*. As well, the UN publishes an annual Human Development Report with the goal of "putting people back in the centre of the development process." Its 2006 report was devoted (for the first time) to the world water crisis. *Beyond Scarcity: Power, Poverty and the Global Water Crisis* cites a form of "water apartheid" that divides those with access to too much clean water and those with little or no access at all.

The United Nations Environment Program monitors water quality around the world and publishes research on a country-by-country basis. UNESCO's International Hydrological Programme has launched a massive water database, headed up by the renowned Russian scientist Igor A. Shiklominov, called World Water Resources and Their Use. The database, which is constantly updated, contains information on the allocation of the world's water resources as well as world water use and availability. UNESCO also published a 2007 report on groundwater called *Groundwater in International Law: Compilations of Treaties and Other Legal Instruments*. The UN Food and Agriculture Organization also released a report in 2006 called *Water for Good, Water for Life: Insights from the Comprehensive Assessment of Water Management in Agriculture*, which contains the work of more than four hundred hydrologists, agronomists and other scientists. In

May 2007, the UNDP Intergovernmental Panel on Climate Change released a *Technical Paper on Climate Change and Water* for expert review by governments around the world.

Every two years, the Pacific Institute for Studies in Development, Environment and Security, led by the noted water expert Peter Gleick, publishes a comprehensive study called *The World's Water: The Biennial Report on Freshwater Resources*, with massive data on the most significant trends in water and water management. The Pacific Institute maintains a website devoted to continuous publication of new information and studies on all aspects of the world water crisis.

Sandra Postel's Global Water Policy Project is dedicated to the preservation of the world's water resources and puts out a steady stream of excellent research and documentation, especially on the desertification of the planet. The World Watch Institute, whose website declares that "Water scarcity may be the most underappreciated global environmental challenge of our time," has an extensive water program and turns out huge volumes of research on the state of the world's water. The International Rivers Network is the best source on large dams and their consequences. In 2007, a number of German universities, in conjunction with the German Ministry of the Environment, published a compilation of new research by more than one hundred international scientists called *Global Challenge: Enough Water for All?* The collection is a comprehensive and compelling collection that links climate change with the planet's growing water scarcity.

The Chinese Academy of Scientists – the country's top scientific body – publishes volumes on the melting glaciers of the Tibetan Plateau and the groundwater crisis in China. Tushaar Shah of the International Water Management Institute's station in Gujarat, India, publishes detailed information about overpumping of groundwater in Asia, warning the world of the "coming anarchy" if controls are not set. The Water and Sanitation division of the Pan American Health Organization monitors water quality in Latin America and produces volumes of research on this region. The U.S. Geological Survey, the Environmental Protection

Agency and the National Academies of Science have, among others, chronicled the growing water crisis in the United States.

Several books were helpful on describing the crisis, including *The Atlas of Water*, a book of facts and maps about water by Robin Clarke and Jannet King (2004); *Deep Water*, on the global struggle against large dams, by Jacques Leslie (2005); *Liquid Assets*, on the need to protect freshwater ecosystems, by Sandra Postel (2005); the June 2007 report *Making Water: Desalination – Option or Distraction for a Thirsty World* by the WWF; *Mirage: Florida, and the Vanishing Water of the Eastern United States* by Cynthia Barnett (2007); and *When the Rivers Run Dry*, on the ecological water crisis, by Fred Pearce (2006). "The Rise of Big Water" by Charles C. Mann in the May 2007 edition of *Vanity Fair* is an excellent source on China's water crisis.

Chapter 2: Setting the Stage for Corporate Control of Water

Several books were helpful references for this chapter chronicling the history of the campaign to enforce a private model of water delivery and its failure. These include *Water Wars* by Vandana Shiva (2001); *Whose Water Is It?*, a 2003 collection edited by Bernadette McDonald and Douglas Jehl for the *National Geographic; The Water Barons* by the International Consortium of Investigative Journalists (2003); and *The Water Business* by Ann-Christin Holland (2005).

I am indebted to Professor Michael Goldman, McKnight Presidential Fellow with the University of Minnesota's Department of Sociology and Institute for Global Studies, for his May 2006 paper, *How "Water for All!" Policy Became Hegemonic: The Power of the World Bank and Its Transnational Policy Networks* presented to the Science, Knowledge Communities and Environmental Governance conference at Rutgers University. Volunteer researcher Tiffany Vogel was helpful in sorting through these studies.

Many Public Services International (PSI) and Public Services International Research Unit (PSIRU) reports and studies found their way into this book, and I am deeply grateful to David Boyes and David Hall and their teams. I am also indebted to the World Development Movement (WDM) and Corporate Europe

Observatory (CEO), who collaborated with PSI and PSIRU on a number of studies. These include *Water Multinationals: No Longer Business as Usual* by David Hall (March 2003); *Water Finance: A Discussion Note* by David Hall (January 2004); *AquaFed: Another Pressure Group for Private Water* by David Hall and Olivier Hoedeman (March 2006); *Pipe Dreams: The Failure of the Private Sector to Invest in Water Services in Developing Countries* by WDM and PSIRU (March 2006); *Down the Drain: How Aid for Water Sector Reform Could be Better Spent* by FIVAS and WDI (November 2006); *Murky Water: PPIAF, PSEEF and Other Examples of EU Aid Promoting Water Privatization*, CEO (March 2007); and *Water as a Public Service* by David Hall and Emanuele Lobina (March 2007).

Other good sources are *Corporate Hijack of Water: How World Bank, IMF, and GATS-WTO Rules Are Forcing Water Privatization* by Vandana Shiva, Radha Holla Bhar, Afsar H. Jafri and Kunwar Jalees (December 2002); *Privatization of Water, Public-Private Partnerships: Do They Deliver to the Poor?* by the Norwegian Forum for Environment and Development (April 2006); *Privatization in Deep Water? Water Governance and Options for Development Cooperation* by Annabelle Houdret and Miriam Shabafrouz, for the Institute for Development and Peace at the University of Duisburg-Essen (2006); and *Privatizing Basic Utilities in Sub-Saharan Africa: The MDG Impact* by Kate Bayliss and Terry McKinley for the UNDP International Poverty Centre (January 2007).

Washington-based Food and Water Watch has also provided excellent research material, including *Will the World Bank Back Down? Water Privatization in a Climate of Global Protest* (April 2004); *Going Thirsty: The Inter-American Development Bank and the Politics of Water* (March 2007); and *Challenging Corporate Investor Rule: How the World Bank's Investment Court, Free Trade Agreements and Bilateral Investment Treaties Have Unleashed a New Era of Corporate Power* by Sara Grusky of Food and Water Watch and Sarah Anderson of the Institute for Policy Studies (April 2007).

Chapter 3: The Water Hunters Move In

Sources for this chapter on the growth of the new sectors in the global water industry were harder to come by as groups, research institutes and universities to this point have undertaken very little research. Sources include company websites and reports, as well as the Fortune 500 Index. Sources also include news reports, which are cited in the chapter. Also useful were the websites, journals and annual reports of industry groups, such as the *International Journal of Nuclear Desalination, Environmental Business International* and *Water Industry News*. I researched the stock market to find out how extensive the water business has become – unfamiliar territory for me.

Global Water Intelligence (GWI) is a monthly electronic newsletter providing up-to-date information on the water industry and was an invaluable source for this chapter. It is very expensive to subscribe to and I had to depend on friends for access to it. GWI also releases reports on key subjects, which are used here. These include *Water Reuse Markets, 2005–2015, Desalination Markets 2007* and *Desalination in China 2007*. Another industry source is *Masons Water Yearbook*, published every two years by the U.K.–based international commercial water infrastructure law firm of Pinsent Mason and considered the bible of the water industry.

Australian author and environmentalist John Archer was very helpful with his critique of water technology in his 2005 book, *Twenty-Thirst Century: The Future of Water in Australia*. A number of organizations are conducting research and campaigns on bottled water. They include the Earth Policy Institute; Corporate Accountability International; War on Want; Campaign to Stop Killer Coke; the Polaris Institute; and the India Resource Centre. Organizations with a critical analysis of nanotechnology include Friends of the Earth International; the ETC Group; Greenpeace; the International Center for Technology Assessment; the British Royal Society; and the Natural Resource Defense Council.

Chapter 4: The Water Warriors Fight Back

My major sources for this chapter on the global water justice movement are the members of the movement itself and my own personal connections to them and their struggles. I have had the privilege of traveling to most of the countries and communities described here and so was able to go straight to the players themselves to verify facts and get stories. I was lucky to have some great resources in Anil Naidoo of the Blue Planet Project and Sara Grusky and Maj Fiil Flynn of Food and Water Watch, who keep up-to-date with all the campaigns, both in person and with extensive web data. Similarly, Public Services International keeps track of the individual privatization fights around the world and its site was very helpful.

Several books were excellent resources for this chapter. They include *Agua Para Todos* by Dieter Wartchow, formerly head of Corsan, the public water company of Porto Alegre Brazil (2003); *Cochabamba! Water War in Bolivia* by Oscar Olivera (2004); and *Reclaiming Public Water* by Corporate Europe Observatory and the Transnational Institute (2005). Also very helpful on the Bolivia situation is Jim Shultz's Democracy Center website. His April 2005 report, *Deadly Consequences: The International Monetary Fund and Bolivia's "Black February,"* gives historic background to the eventual win in that country.

Other helpful reports were *Untapped Connections: Gender, Water and Poverty* by WEDO, the Women's Environment and Development Organization (2003); *Ganga: Common Heritage or Corporate Commodity?* by Vandana Shiva and Kunwar Jalees (2003); *Water Justice for All: Global and Local Resistance to the Control and Commodification of Water* by Friends of the Earth International (January 2003); *Water for the People* by IBON in the Philippines (November 2004); *Taking Stock of Water Privatization in the Philippines: The Case of the Metropolitan Waterworks and Sewage System* by the Freedom from Debt Coalition and Jubilee South of Asia-Pacific (December 2004); *Commercialization and Privatization of the Indonesian Water Resources* by the Indonesian Forum on Globalization (2004); *Drinking Water Crisis in Pakistan and the Issue of Bottled Water: The Case of Nestlé's "Pure Life"* by the Swiss Coalition of Development Organizations (April 2005); *Faulty Pipes: Why Public Funding – Not Privatization – Is the Answer for U.S. Water Systems*, Food and Water Watch (June 2006); and

Where Does It Start? Where Will It End? Las Vegas and the Groundwater Development Project by PLAN, the Progressive Leadership Alliance of Nevada (January 2007).

Chapter 5: The Future of Water

There are many important sources for this chapter on the need for a water covenant and the right to water. On water as a national security issue in the United States, two reports were particularly helpful: *Terrorism and Security Issues Facing the Water Infrastructure Sector* by Claudia Copeland and Betsy Cody of the Congressional Research Service (January 2005) and *Global Water Futures: Addressing Our Global Water Future* by the Center for Strategic and International Studies and Sandia Laboratories (September 2005).

On the issue of water conservation and watershed protection, the work of Sandra Postel and Peter Gleick and their respective institutes is invaluable. David Schindler of the University of Alberta has done groundbreaking research on freshwater protection, and Rajendra Singh and his organization, Tarun Bharat Sangh, has taught villagers all over Rajasthan to take charge of their water resources and harvest rain for sustainable farming. *Who Owns the Water?*, a 2007 anthology edited by Klaus Lanz, Christian Rentsch, Réne Schwarzenbach and Lars Muller, is a good source of ideas. The chapter entitled "Water: A Shared Responsibility" in Bill McKibben's 1990 bestseller *The End of Nature* lays out a detailed plan of action to conserve and protect source water. Michal Kravçik has written extensively on his concerns about the hydrologic cycle and how to protect it. He lays out a plan in *Blue Alternative: Water for the Third Millennium* (2002).

On public alternatives to private water control, Public Services International was once again a very valuable source. So are the following reports: *In the Public Interest: Health, Education, and Water and Sanitation for All* by Oxfam International and WaterAid (2006); *Public Water for All: The Role of Public-Public Partnerships* by the Transnational Institute and Corporate Europe Observatory (March 2006); *Going Public: Southern Solutions to the Global Water Crisis* by the World Development Movement (March

2007); and *Down the Drain: How Aid for Water Sector Reform Could be Better Spent* by FIVAS and World Development Movement (November 2006).

There is a wealth of good material on the right to water and a UN convention. Ashfaq Khalfan of the Right to Water Program at the UN Centre for Housing Rights and Evictions (COHRE) has written extensively on the issue. In March 2004, he wrote *Legal Resources for the Right to Water: International and National Standards.* So has Rosmarie Bar of the Swiss Coalition of Development Organizations, as in her January 2004 publication, *Why We Need an International Water Convention.* Bread for the World and the Heinrich Böll Foundation teamed up with COHRE in a March 2005 Global Issues Paper called *Monitoring Implementation of the Right to Water: A Framework for Developing Indicators.* John Scanlon, Angela Cassar and Noemi Nemes of the World Conservation Union wrote *Water as a Human Right?* on the legal ramifications of a UN instrument. Henri Smets of the European Council on Environmental Law and the French Academy of Water compiled a catalogue of all current domestic legislation in his 2006 report, *The Right to Water in National Legislatures.* Rodrigo Gutiérrez Rivas of the Legal Resource Institute at the University of Mexico wrote a March 2007 paper called *Privatization and the Right to Water: A View from the South.*

As always, Steven Shrybman has been an excellent source. He wrote *A Critical Review of the "Green Cross" Proposal for a Global Framework Convention on the Right to Water* in 2005 and an April 2007 submission to the High Commissioner for Human Rights for the Council of Canadians found on the council website, www.canadians.org.

Two important new films on the global water crisis tell this story, and I highly recommend them: *For the Love of Water* directed by Irena Salina and produced by Steven Starr (released fall 2007); and *Blue Gold* directed by Samuel Bozzo (to come out in 2008).

Index